OTHER MINDS

SYNTHESE LIBRARY

STUDIES IN EPISTEMOLOGY,

LOGIC, METHODOLOGY, AND PHILOSOPHY OF SCIENCE

VOLUME 246

ALEC HYSLOP
La Trobe University, Melbourne

OTHER MINDS

KLUWER ACADEMIC PUBLISHERS
DORDRECHT / BOSTON / LONDON

Library of Congress Cataloging-in-Publication Data

Hyslop, Alec.
Other minds / Alec Hyslop.
p. cm. -- (Synthese library ; v. 246)
Includes bibliographical references (p.) and index.
ISBN 0-7923-3245-8 (HB : alk. paper)
1. Other minds (Theory of knowledge) I. Title. II. Series.
BD213.H97 1995
121'.2--dc20 94-41364

ISBN 0-7923-3245-8

Published by Kluwer Academic Publishers,
P.O. Box 17, 3300 AA Dordrecht, The Netherlands.

Kluwer Academic Publishers incorporates
the publishing programmes of
D. Reidel, Martinus Nijhoff, Dr W. Junk and MTP Press.

Sold and distributed in the U.S.A. and Canada
by Kluwer Academic Publishers,
101 Philip Drive, Norwell, MA 02061, U.S.A.

In all other countries, sold and distributed
by Kluwer Academic Publishers Group,
P.O. Box 322, 3300 AH Dordrecht, The Netherlands.

Printed on acid-free paper

Printed in the Netherlands

TABLE OF CONTENTS

Preface		vii
Introduction		1
1.	What is the problem	5
2.	Who has the problem of other minds?	15
3.	Other minds and scientific inference	29
4.	The analogical inference to other minds	41
5.	Criteria and other minds	71
6.	The "old" private language argument and other minds	93
7.	The "new" private language argument: Kripstein's new form of scepticism	105
8.	"Wittgenstein's" "attitudinal approach" to other minds	121
9.	Strawson on other minds	131
Notes		141
References		151
Index		155

PREFACE

This book has been a long time in the making. Other issues have taken me away from it from time to extended time. But I kept coming back to the problem of other minds. It has remained a great issue, it is much contested still, and it is, after all, close to us all. I like believing that the time taken has deepened my understanding of the problem and how it is to be handled.

Other people, some by disagreeing vehemently, have helped greatly. I mention in particular, Brian Ellis, Robert Fox, Graeme Marshall, Tim Oakley, Ray Pinkerton and Robert Young. Robert Pargetter argued with me, and kept insisting that I write this book. John Bigelow, Michael Bradley, Keith Campbell, Frank Jackson, and William Lycan assisted by reading an earlier version and providing valued comments. Frank Jackson has been specially helpful, not just on this topic. He can be blamed for initially causing me to take the analogical inference seriously. The La Trobe Philosophy Department has been a good place to do philosophy.

I am grateful to Suzanne Hayster, Sandra Paul, and Betty Pritchard for struggling at various times with various recalcitrant manuscripts. Most particularly I thank Gai Larkin. She has seen the project through, with considerably more than efficiency.

Finally, Alison, Michael, Sandy, and Zoe have profoundly enriched both me and my understanding of other minds. My father provided an enduring philosophical example for me, though I have only recently come to recognise this. I end with my thanks to Diane. She has not allowed philosophy to come between us.

INTRODUCTION

> "To return to Françoise ... if then in my anger at the thought of being pitied by her I tried to pretend that on the contrary I had scored a distinct success, my lies broke feebly on the wall of her respectful but obvious unbelief. . . . For she knew the truth." - Proust.

> "... it was she who first gave me the idea that a person does not (as I had imagined) stand motionless and clear before our eyes with his merits, his defects, his plans, his intentions with regard to ourself exposed on his surface, like a garden at which, with all its borders spread out before us, we gaze through a railing, but is a shadow, which we can never succeed in penetrating, of which there can be no such thing as direct knowledge, with respect to which we form countless beliefs, based upon his words and sometimes upon his actions, though neither word nor actions can give us anything but inadequate and as it proves contradictory information - a shadow behind which we can alternately imagine, with equal justification, that there burns the flame of hatred and of love." - Proust.[1]

Most people find at least some other people mysterious. Some people find at least some others transparent. Some people are perhaps mysterious to some and transparent to others. Some people find all others mysterious. Perhaps there are some who find all other people transparently on view, though that seems (significantly) much more doubtful. Most of us, however, find that some people are more difficult than others to understand and people in general are more difficult to understand at some times than at others. Mostly, we get by well enough, sometimes regretting, sometimes welcoming, the degree of difficulty that attends the business of fathoming the actions of other people.

In all this it is not doubted though, that there are, indeed, other people, be they varyingly mysterious, transparent, or whatever. Outside of professional scepticism or psychopathology, we are all sure that we are not alone. Other human figures are, at least to an extent, sufficiently as we are. We are sure they too engage in thinking, reasoning, doubting and believing. We are sure they feel sick, feel pain, feel happy and so on, from time to time.

But what entitles us to such sureness? How is our belief that we are not alone justified? Here sureness gives way to uncertainty. There is no generally accepted answer. It is not even clear that, generally, people have their own answer. Our confidence seems undisturbed. Neither the lack of general agreement nor the absence of a personal view shakes our general sureness.

This sureness both suggests that we are, indeed, justified in our sureness and demands that we need to be justified, personally, each of us. Whether or not the answer has been articulated, by anyone, it needs to be available to us all, in principle, and "internalised" by each of us, in fact. It must give us a secure ground for our believing in others, sustain the sureness. It need not, though, be a matter of its having caused us to believe, initially. The answer needs to be what keeps us going, not what got us started. The available answers have to meet this test, sustain the sureness.

It would, after all, be extraordinary, and fascinating, if there were no justification (and this could be demonstrated). Some would be terrified, some, perhaps, exhilarated. Some philosophers think it would make no difference. We would go on undisturbed. Much more interestingly, they think that would be the right thing to do. They even seem to have an argument, of sorts, which is that this extraordinary situation is the actual situation.

How far does our sureness extend? If we are sure that other human beings think and feel, ache and yearn, are we sure they think and feel, ache and yearn as we do? Are their pains as ours are? Is our sunflower their sunflower?

I think most of us think the answer to each of these questions is yes, though not everyone thinks so. I think most people are also sure the answer is yes, if they think it is yes. So we (generally) look for a justification for this particular sureness.

Various justifications are on offer. The traditional analogical inference appealing to the similarity between ourselves and others has long been out of favour. The essential element in that inference is the primacy of one's own case, which makes it a generalisation from one case. The essential thrust of the "scientific inference" approach is to avoid just that primacy and, therefore, that generalisation. That (scientific) approach treats other minds as theoretical entities and seems to be the approach currently, and increasingly, in favour. However, those who support it, though numerous, rarely venture into print.

These two, the old-fashioned analogical inference and the new scientific inference seem to me to be the real choices. The others seem to me to fail. These are the various criterialist views, the private language position (old and new), and Strawson's. It is often claimed that these are significantly different from the previous two, being "dissolutions" of the problems of other minds rather than solutions. That is not true of all of them. But they all of them can be seen as providing a reason for believing that there are other people. What is common to the "failures" is the idea that it is incoherent for each of us to think there is a problem. But if so, a justification for believing in other people is surely that it would be incoherent to believe otherwise. So they can all be viewed as arguments justifying belief in other minds. Of course, nothing follows about their being in fact a justification utilised by us. Indeed, on this view the ordinary person is in fact, justified, given the incoherence of any alternative, while being seen, given that incoherence, as not requiring justification.

I shall be defending the analogical inference as basically right. I shall argue, also, that none of the alternatives work. In particular, treating other minds as theoretical entities, the "scientific inference", does not do the job unless one makes crucial use of one's own case. That is always needed to sustain the beliefs we have about other people.

The alternatives threaten the analogical inference from without. The threat from within is the reliance each of us has on our own case. That has in more recent times seemed to almost all philosophers a fatal defect. Its adherents have tried to extricate themselves by claiming their inference is not, when properly understood, a generalisation from one case. I am sure that it is indeed a generalisation from one case but that it is none the worse for that. The old analogical inference basically remains a good argument. It does the job. So I shall defend resolutely a version of the analogical inference and reject, successively, the alternatives. That inference will be defended as a generalisation from one case. There is no escaping that dependence on our own case.

CHAPTER ONE

WHAT IS THE PROBLEM?

What generates the problem of other minds? Is it that we live and die alone? Some would think that a surprising thought given the number of death-bed scenes in literature. One of the things that might be more plausibly meant by dying alone is that no one else can, by dying with me, share my death. To be less poignant, no one else can share my toothache. But no one else can share my smile, my cover drive, my pitching, my swimming. There is, however, a problem of other toothaches but no problem of other smiles. So living and dying alone, so understood, whether or not a problem, is a different problem.

Is it that we cannot observe the mental states of others? That is not the problem, for two reasons. The first is that though we generally do not observe the mental states of others, it is not clear that it is impossible to do so. Telepathy and various forms of healing might be instances of such observation. They are frequently thought to be.

The second reason is the more important. Were I to observe another's pain, that would not help me with my problem. What would be needed would be my observing it *as* the pain of another. It would need to be "labelled". I should have to know that what I observed was, indeed, being felt by someone else.

Take the case of "Siamese" twins Fred and George who "share" a leg. It comes about that they have a pain in their big toe. Each feels pain. There are various ways to describe the mental typology of these twins with their pain. One way, certainly, is to view the situation as one where they are sharing a pain. So each has a pain, the same pain. Each feels pain, the same pain. Fred feels George's pain and George feels Fred's pain, there being but one pain, which each feels, and each (jointly) owns.

Given our immediate interest, in what generates the other minds problem, it does not matter that there are other ways of classifying this case. Nor does it matter that this particular way be regarded as plausible or implausible. It is coherent and it represents a possible view. It allows us to see that neither the joint ownership of pain, nor the extent to which pains are observable by others, affects in any way the other minds problem.

So what if Fred's pain is one and the same as George's? How is he supposed to know this without knowing that George feels pain (and, in particular, this pain)? And how is he supposed to know that indeed George does? If George does feel pain let us assume that it follows that what George feels is George's pain but is also Fred's, equally. That does not assist Fred in any way to know that George feels pain. His situation is not in any way improved as compared with the normal situation. In feeling what is George's pain, Fred does not thereby learn that what he feels is felt by George.

The same holds for observability. Fred observes, on this view, George's pain, and George observes Fred's. But what Fred needs to observe is that what he feels is felt by George. That it is indeed so is something he needs to establish. It is not given to him in experiencing what is, in fact, felt by George. He needs to experience it not simply as it is being felt by George but as *being felt by* George. So he needs to observe its being felt by George, which is additional to his observing it while George feels it. Fred lacks this direct knowledge *that* George is in pain (while having direct knowledge *of* that pain).

The lack of this direct knowledge does not of itself generate the problem of other minds. That problem is not caused simply by the fact that Fred lacks this direct knowledge of George and George lacks it of Fred, though each does lack it of the other. Fred has this direct knowledge of Fred, and George has it of George, while lacking it of the other. That is the problem. Without the asymmetry, there would be a problem of minds, not other minds. So there is no question that Fred knows he is in pain and George knows he is in pain. They know this directly. They do not infer it. They do not in this way, directly, know this of the other.

So the problem of other minds is generated by an asymmetry in respect of knowledge. Each of us has direct knowledge of our own case but not of others. We know directly that there seems to be a sunflower in front of us and, in particular, how it appears to us, but we do not know directly how it appears to others nor, even, that it appears at all to others. The same holds for toothaches. We do not know about all our mental states directly, however. What we believe, what we feel about this or that, might well require inference on our part. But a good deal of experience is known directly by each of us. Nothing of the experience of others is known in that way.

I have characterised the problem as one generated by an asymmetry in respect of direct knowledge. It cannot be characterised as asymmetrical knowledge, *tout court*, for the analogical arguer (and other arguers to other minds) can hold that they are as sure about how others experience sunflowers as they are about how they themselves do. This sureness is obtained for the analogical arguer by courtesy of the analogical argument, appealing to the similarity that holds between human beings.

There is an asymmetry in that argument, of course, in that my knowledge of other minds epistemically depends on my knowledge of my own but not vice versa. So it would be possible to characterise the other minds problem as just that: knowledge of one's own mind is logically prior to knowledge of other minds. I have not done so for two reasons. Firstly, it is unclear to me how this would be so unless knowledge in one's own case was crucially non-inferential. Otherwise, how could we acquire knowledge of our own mental states without being able to do the same with the mental states of others? Secondly, on this view, treating other minds as theoretical entities, where the best explanation of the behaviour of others is that they have minds would not solve the problem of other minds. It would remove it, for this particular (epistemic) asymmetry is rejected by that approach.

So the problem of other minds is generated by (1) our having direct knowledge of (some of) our own mental states, (2) our not having this direct knowledge of others' mental states. Given this, the various "dissolvings" of the other minds problem can be clearly seen to be just that (if successful) in their different ways. The "private language" approach rejects the claim that we have direct knowledge in our own case independently of any reference to the case of others. The various criterialists reject the asymmetry. We have, according to them, in some sense or other, non-inferential (if not direct) knowledge of other minds. Some of them, indeed, claim that we have direct knowledge of those minds, at least in a sufficient number of instances.

If the problem is generated as suggested, the problem itself is one of justifying our belief in other minds. So conceived, it is then straightforwardly epistemological. As far as I know solipsism is not regarded as an option. It is taken that we believe in other minds. As far as I know, scepticism about other minds is not regarded as an option, other than as part of a more general scepticism. The project is, as it were, to locate the justification.

Tom Nagel, however, has suggested that the interesting problem of other minds is not the one I have outlined but rather that of explaining how we can even have a conception of others:

> The interesting problem of other minds is not the epistemological problem, how I can know that other people are not zombies. It is the conceptual problem, how I can *understand* the attribution of mental states to others.[1]

But how is this a problem? Nagel reminds us that the trick cannot be managed

> by extending the idea of what is immediately felt into other people's bodies, for as Wittgenstein observed, that will only give you an idea of having feelings in their bodies, not of *their* having feelings.[2]

But this rejects a solution without indicating to what it fails to be a solution.

The reference to Wittgenstein is to a famous passage in *Philosophical Investigations*:

> if one is to imagine someone else's pain on the model of one's own, this is none too easy a thing to do: for I have to imagine pain which I *do not feel* on the model of pain which I *do feel*. That is, what I have to do is not simply to make a transition in imagination from one place of pain to another. As, from pain in the hand to pain in the arm. For I am not to imagine that I feel pain in some region of his body. (Which would also be possible.)[3]

Here the problem is constituted by an asymmetry: that between felt and unfelt pain. Nagel seems to endorse this:

> The problem is that other people seem to be part of the external world, and empiricist assumptions about meaning have led various philosophers to the view that our attribution of mental states to others must be analyzed in terms of the behavioral evidence, or as parts of some explanatory theory of what produces observable behavior.

> Unfortunately, this seems to imply that mental attributions do not have the same sense in the first person as in the third.[4]

Presumably this is the same problem: I do not feel other people's pain. Nagel's way of putting it is helpful in clarifying the connection between this (conceptual) problem and the epistemological problem. The latter looks for the justification of our belief in the pains of others given that we cannot feel them. The former wonders how we can conceive of those pains, given that we cannot feel them.

The problem is not, however, that of moving from pain that is felt to pain that is not felt. Each of us has that problem internally. If I remember being in pain, or fear being in pain, then I have in mind pain I *do not feel*. The problem exercising Wittgenstein is surely that of moving from pains being felt by me to a conception of pains being felt by someone else. Despite what he writes, whether or not I feel what they feel is irrelevant, as I have argued previously, in the context of the epistemological issue. Here the same point emerges. First, from the fact that each of us remembers pain. So the point is not whether pain is felt or not. Secondly, from the fact that we might feel pain in another's body. This is not what is needed. Wittgenstein is perhaps more likely to have been pointing out that this would be a case of my having a pain in someone else's body. But he could have been drawing attention to the fact that what is needed is the concept of someone else's being in pain, which is distinct from my having a pain in their body, and from my feeling what is in fact also felt by them.

Why is the problem as stated, a problem? Nagel's explicit explanation of its being so depends on a particular (unattractive) theory of meaning, the empiricist theory of meaning. Otherwise it becomes a question rather than a problem: how do we get the relevant concept? What makes it a problem is surely once again the asymmetry. Each of us has direct knowledge of our pains but not of the pains of others. That is what generates the epistemological problem. That is what generates the conceptual problem. Feeling another's pain would not help unless it was felt as another's. It is the asymmetry that is behind both Wittgenstein's and Nagel's setting of the "problem". This is seen in Wittgenstein's case in the fact that my remembered pains are "modelled" on pains I feel though they are not, obviously, pains I feel. So his real concern is different. In Nagel's case, the empiricist theory of meaning notoriously struggles to avoid giving two different accounts of pain, one first-person and the other third-person.

So the conceptual problem (given the asymmetry) is to acquire the notion of a person's feeling pain. That perhaps sounds strange since each of us is a person. Talk of a person's feeling pain conceals the asymmetry. Having the concept of a person's feeling pain involves oneself and others. Even here, it is tempting (with Nagel and Wittgenstein) to see the problem in terms of feeling pain. One needs to feel pain to have the concept. But though true (even if not a necessary truth) that is not the conceptual problem. If I felt pain by feeling another's pain then I should have satisfied that condition. Feeling pain is one thing, having the concept of a person's feeling pain is another.

Looked at another way, if the persons did not include oneself, then there would be no unusual problem. One would make what headway one could, from the outside. There would be no asymmetry. That is introduced by introducing oneself. How much of a problem is then created is a matter of how easily we can show how the relevant concept can be acquired. The task is to show how we can have the concept of other persons' feeling pain as well as ourselves, given our now familiar asymmetry.

Our own remembered and anticipated pains ease us through. They are cases of our having felt pain, or anticipating feeling it, but they do not involve feeling pain as we remember, or anticipate. In the case of memory, I think of a person, myself, feeling pain. I can then shift to imagining a person, myself, feeling pain. The shift from doing that to imagining another person feeling pain is manageable. Take an act of visualising oneself in pain, which is a specific instance of imagining oneself in pain. This effects, in stages, the desiderated "distancing". One visualises oneself in pain. But one does not (strictly, need not) while in pain visualise oneself in pain. Not in pain, one visualises oneself in pain. One visualises oneself from the outside, in pain. One visualises oneself as one would another, in pain. Oneself as another can then give way to visualising another, in pain.

Imagine a "line-up", which includes oneself. If one can visualise oneself in that "line-up", in pain, then the detachment involved, between oneself as current imaginer and oneself in the line-up, allows one to visualise another in that line-up in pain. We could hardly do one without the other. One is not visualising another while imagining one feels pain in that other body; just as one is not imagining one's present pain being felt in an earlier self. One is visualising one's earlier (other) self in pain. To imagine that one is now feeling pain in that earlier self's body is a distinct act (one might carry it out by imagining the earlier body has been somehow preserved).

Another approach to the required conception of others' feeling pain is to begin with other human bodies and envisage oneself without experiences of pain. One watches these bodies and wonders. They exhibit, occasionally, pain-behaviour. One theorises about inner episodes responsible for the behaviour one observes. Then, suddenly, one feels pain and that pain fits with one's theorising about inner episodes in others. Here we have the birth of the full-fledged concept of the inner episode that is pain. There seems no temptation to see any problem here. Of course, one might adopt the view that whatever the inner episode is in each case, provided it is responsible for the behaviour, it is pain. But one might believe that it is the same in each case, given the similarity of behaviour, and wonder what it is. It is for this latter view that the concept waits on the individual experience. The former view does not wait in this way, and any differences between the individual cases would be irrelevant to the concept of pain.

McGinn[5] seems to reject the way through imagination to the desiderated conception. He draws an analogy between visual squareness and tactual squareness on the one hand, and first-person pain and third-person pain on the other. In each case he claims there is a mode of presentation. One needs to abstract from the mode of presentation to arrive at, respectively, the (objective) concepts of squareness and of pain.

McGinn sees a problem in the case of pain. Getting to third-person pain through imagining first-person pain (and vice versa) is akin to trying to get to tactual squareness through imagining visual squareness (and vice versa).

But there is a crucial difference. Squareness is (indifferently) presented either way. Pain is *only presented* in the first person. The way through imagination to other pains recognises this asymmetry. In the case of pain, imagining first-person pain on the basis of third-person pain is straightforward if the inner episode is denied and would be straightforward if it were presented (as an inner episode) in both cases. It is only problematic *if* an asymmetry is accepted.

McGinn claims that we need "to prescind from the sensory modes in which the property of squareness is presented"[6] and equally for pains. But this is just what we do, *insofar as we need to*, in the steps outlined previously, moving from my being in pain, to remembering being in pain, and so on. We do *not* prescind from pain being something *felt* (though we do prescind from its being something felt by me alone). In that sense, the "sensory" mode is ineluctable. But the move from first-person to third-person through imagination is precisely not analogous to the move from visual awareness to tactual awareness.

McGinn thinks we can prescind in the case of squareness but not in the case of pains. Were he to mean that *someone* feels pain then that would be true but no matter. McGinn means, however, that in imagining a person in pain I must either do it from *my* perspective (first-person) or from another. If it is the latter then on his view I do so by (say) visualising another *but limit myself to what I can visualise.*

The parallel depends on the limitation. I imagine that I am in pain. I imagine she is in pain. I imagine (visually) squareness. I imagine (tactually) squareness. The difference in sensory mode depends on taking her pain as being presented externally. But there is no mode in which pain is presented other than to me. If we ignore that, we have the required parallelism. Given the limitation, I shall be defeated if I try to imagine third-person pain by imagining first-person pain. One is presented internally, the other is "presented" externally. In one case I imagine pain, in the other I imagine "pain" (really, behaviour).

Ignoring the limitation, it is not true that my (imaginative) thought is constrained as he claims. It need not be one or the other person. If I am asked to imagine a person in burning pain then that is different from being asked to imagine a burning pain (I can visualise a person in burning pain but I cannot visualise burning pain). Imagining burning pain can be indifferently realised by my imagining Ronald Reagan feeling it, my feeling it, a chimpanzee feeling it, no one in particular feeling it (though it has to be felt). If we compare this with the line-up case, there might be no answer as to whether it is I or someone else who feels it. This is not the issue. It is one thing to imagine yourself in pain, another to imagine another person in pain, another to be asked to imagine a person in pain, another a creature in pain, another a pain.

The crucial point, however, is that McGinn seems to be taking it that one approaches pain either from the inside or from the outside and if one approaches it from the inside one is limited to the first-person case. That is not so, and the break-out is, as previously outlined, through remembered pain where pain is remembered but not felt. One combines the first-person and third-person approach in imagining oneself in pain where one does so by visualising oneself in pain. Then one does the same, with another.

There is an objection from Christopher Peacocke[7] to the way through imagination. He points out that images cannot do the job. The image needs to be set in a conceptual context. So an image of my pain cannot in itself be distinguished from an image of your pain. Each is conceived of as distinct, so the required concepts are needed in order to imagine appropriately. They cannot, therefore, be obtained through imagination.

It is true that images alone cannot do the job. However, it is not true that the imaginative steps presuppose the concepts required. Take the case of my imagining that I was in pain yesterday in my room. I visualise myself as I think, in my room, in pain. Unfortunately, for the moment I mistakenly think my body looks other than it does; indeed, that it looks just as my neighbour's looks. Suddenly, I realise my error, and realise that I was imagining someone else in pain. I was not imagining that I felt pain in someone else's body. I thought that was my body. I took myself to be imagining that I was in pain but I failed, because, in effect, I was imagining the owner of that body in pain. As soon as I realise my error I am equipped with the appropriate concept, via my error.

If the worry is expressed that one is stuck with one's own case, that one (necessarily) learns from one's own case (on this view), then one thinks of Malcolm and the view that it would follow that pains are felt only by oneself. That unpromising line of argument is distinct from the present concern (and, anyway, has been despatched by Plantinga).[8] Malcolm's essential thought was that if I learn what pains are from my pains, since those pains are necessarily mine, I would learn that pains are necessarily mine. That would be my concept of pain. The essential response is that my smiles are necessarily my smiles. So the same problem would exist with other smiles. There is no such problem. It follows there is none with other pains (at any rate, not from this source).

So we are left with the epistemological problem of other minds. I have claimed that problem is generated by (1) our having direct knowledge of (some of) our own mental states, and (2) our not having this direct knowledge of (any of) the mental states of others.

We do not, of course, have direct knowledge of all of our mental states. There is no commitment here to the "self-intimation" thesis, to our mental lives being transparent to ourselves. However, within each broad category of the mental, direct knowledge can exist and does exist. Sometimes at least we know directly what we think, and feel, and why we do what we do. Extensively, we know directly about our pains and other bodily sensations, about our sensory experience. None of this is true of others.

There is no commitment in this view to the "incorrigibility thesis", that we cannot be mistaken in our beliefs about our mental states. The direct knowledge being claimed is not being claimed to be infallible. It is being claimed to be knowledge. That is what is needed, not infallibility.

What is crucial, it should be noted, is that there is an asymmetry. That is what matters rather than what view is taken of it. If in the no-nonsense spirit of cognitive science one talks of self-scanning or of inner perception as opposed to external perception, that still leaves the required asymmetry. Self-scanning is introspection, providing knowledge directly. Other brains when "scanned" do not provide symmetrical information. We are left with the problem of other minds as long as we have the crucial asymmetry.[9]

CHAPTER TWO

WHO HAS THE PROBLEM OF OTHER MINDS?

Who has the other minds problem? Are there views of the human mind that avoid the problem? Those who believe that people are more than complex (physico-chemical) machines, are more than complex aggregates of sub-atomic particles which have only material properties, have non-material properties as well as material properties, certainly have the problem. So all Dualist views of the nature of human beings have the problem. So clearly do Idealist views, believing there are only minds and their ideas.

What about Materialists? There are different kinds of Materialism. It might be thought that Eliminativists, who deny there are mental states, avoid the problem: if there are no mental states then there are no other minds to worry about. But how is it known that there are no minds? I shall return to this.

Those Materialists who think mental states are nothing over and above brain states nevertheless think mental states exist. Such (non-eliminativist) Materialists separate into two groups. One offers an account of the (conceptual) nature of mental states preparatory to affirming that they are, in fact, brain states (and have only material properties). Such accounts are generally Functionalist. The other group offers no account of mental states distinct from that offered by Dualists (presumably they are itemised by ostension generally) and differs from Dualists precisely in affirming that mental states are purely material. On this view each mental state is a material state, though not every material state is a mental state. Nor is it being claimed that every material state of the same kind as those being mental states will also be a mental state. So far, this brand of Materialism asserts *only* that each particular mental state is a material state (with only material properties).

This second group of Materialists straightforwardly shares the other minds problem with Dualists. Their claim is that mental states, as outlined, are, in fact, brain states. Who has them is a separate issue. That human beings generally have mental states in no way follows from the claim that mental states are brain states. If mental states are brain states and brain states are observable then mental states are observable. But they are not, thereby, observably mental states. If in observing a brain state I am in fact observing a mental state, it fails to be the case that in doing so I am sure, from doing so, that I am observing a mental state. That is what would be needed, not merely observability. The brain states would have to come labelled, as mental states. So this brand of (homespun) Materialism does not avoid the other minds problem.

What of the other brand of Materialism, which offers an explicit account of the nature of the mental as an integral part of its case for Materialism? Behaviourism was the first such attempt and its spirit remains strong. In its simplest form, being in a certain mental state is entirely a matter of behaving in a certain way (in certain circumstances). Such a view would seem to banish the "inner life". Everything would be "upfront". Pain would be observable. The problem of other minds would vanish. Or so it seems: I shall return to this, also.

Of course, such Behaviourism, variously dubbed "Simple", "Naive" or whatever, will not do, so its success or otherwise with other minds seems academic. Its more sophisticated successor, Dispositional Behaviourism, though thoroughly unpopular, is more respectable. However, its sophistication, its viewing mental states as involving in addition to actual behaviour dispositions to behaviour, would seem to run the risk of reviving the other minds problem. Behaviour is observable whereas dispositions are observable only when exercised. I can know directly that I am disposed to behave in a certain way without doing so but others lack this direct knowledge. So the problem might get off the ground. However there would seem to be sufficient exercising of dispositions to behave for adequate grounding to be provided for the belief that other human figures have minds. Even the more torpid among us behave sufficiently often in sufficiently varied ways.

So both forms of Behaviourism seem to avoid the other minds problem, though neither is popular nowadays and they would only be of use if it is at least reasonable to believe that they are true. I think both fail that test. However, I shall return to both, as well as to Eliminative Materialism.

Behaviourism, though no longer an option, for philosophers at any rate, did not fall to argument, good or bad. It was superseded by other forms of Materialism, more specifically by Central State Materialism. Behaviourism sees having a mind as purely a matter of an organism's behaving in certain ways and being disposed to do so. Its mistake is locating the mental where it does, externally. Whatever the mental is it is internal, within the organism. It mediates between the organism and its environment. It is not behaviour. It causes behaviour. As a general truth, few would disagree. For Central State Materialism, however, that is all there is to the mental. The mental is that which mediates, which gives rise to behaviour. Behaviourism has seeped through to the inside. The mental is not behaviour but is what gives rise to behaviour. But what makes an internal state of the organism mental is a matter of behaviour. What differentiates mental states from one another is a matter of the different behaviour they respectively cause. Behaviourism is swallowed almost whole but seems undigested. If we think of the case of pain then it has been, rightly, moved inside but the hurtfulness seems still homeless.

With Central State Materialism, we seem to have an other minds problem, though it is (suspiciously) easy to solve. We have the problem because we have the relevant asymmetry, provided we allow that we have direct knowledge of the relevant internal states in our own case. That seems to be universally allowed by Central State Materialists. In Armstrong's terms, where mental states are treated as (generally) states apt for the production of behaviour, he allows we can know we are in such states without their giving rise to behaviour. As for other minds, where we lack this facility, we deal with them thus, faced with the behaviour of others:

> We need only three premises to infer the existence of a mind that this behaviour is expressive of (i) the behaviour has *some* cause; (ii) the cause lies in the behaving person; (iii) the cause is an "adequate" cause - it has a complexity that corresponds to the complexity of behaviour. Given only these quite modest assumptions, the existence of another mind is necessary.[1]

The thrust seems acceptable. So the problem exists but seems to be speedily handled. However, I shall return to this, too.

Finally, we reach Central State Materialism's offspring, Functionalism. The Materialism has slipped off (only to the wings, of course) and it is non-speciesist. The mental is a matter of there being internal states having a certain role. Alien beings, computers, ghosts or whatever can have minds if they satisfy this test, by having internal states occupying the relevant roles. The internal states might in fact be states of a brain, of a computer, of a soul, but that is irrelevant to their status as mental states. It does not matter what in fact occupies the role.

In the case of human beings, the roles in general are filled by brain states which mediate between human beings and their environment. So the differences between Central State Materialism and Functionalism are irrelevant to the other minds problem. Each has a problem but the problem has an easy solution. What is crucial and produces this result in each case is that what constitutes the mental is a matter of there being internal states with a certain role and that is all there is to it. Whether or not qualia exist is not to the point. It is the view taken of them that matters. Functionalists bearing qualia perhaps are to be distrusted but their qualia will be the occupiers of roles. What other properties qualia might have will be irrelevant to their status as instances of the mental. Qualia will only be thought to bring the classical (and difficult) other minds problem if that condition is abandoned and they are permitted to have mental properties other than the roles they play. So the careful Functionalist can embrace even the inverted spectrum.

We seem to be left with an untidy picture. Behaviourism and Eliminative Materialism have no other minds problem. Central State Materialism and Functionalism have the problem but have a comparatively easy solution involving straightforward (scientific) inference from what can be observed to what cannot be observed. Dualists and Idealists have the other minds problem and do not seem to have available the Central State Materialist and Functionalist solutions since they think there is more to the mental than the occupying of certain roles. The hurtfulness of pain is mental and not reducible to a role or roles.

Those who are against untidiness and those who are against only unnecessary untidiness will both be relieved to know that the true picture is tidy. All theories of mind have the same other minds problem. None has a solution available to it easier than any other.

It has sometimes been suggested, jokingly, that Behaviourists are different. They do not feel pain. That is why they are Behaviourists. That is why Behaviourism is satisfactory for them. The rest of us do feel pain. That is why it is not satisfactory for us. It is not that they are leaving something out. They have had something left out. The joke is philosophically deep. It is possible that Behaviourists are different. That is not of course a reason for thinking that they are indeed different. It is possible, after all, that they are the same. We laugh because we are sure they are not (laughter otherwise would be uneasy). Behaviourists share this confidence. They believe they are the same as others. They believe that what holds for them holds for others, that it is with others as it is with themselves. Behaviourism is not to be true only for Behaviourists. It is thought, by them, to be true. It holds for human beings. It is not, in particular, thought by Behaviourists to be true, only of themselves. It is not offered as a possibility for others, known to be true only in the case of oneself.

Behaviourism is a theory of mind. It is not a theory holding only of the propounder's mind (nor is it a theory only of minds other than the propounder's). It is supposed to cover all human beings. It follows that it cannot be used to establish that there are, in fact, other minds. The Behaviourist cannot argue: Behaviourism holds for me, therefore it is true of others, therefore there are other minds (by various easy steps). This is blatantly question-begging. Only if it holds for others can it be used to show there are others.

Take what is perhaps the simplest case, the Simple Behaviourist. Either she is different from others or she is the same. If she is the same then the asymmetry in direct knowledge applies. She has direct knowledge of herself, not of others. She has to get from what she believes to be true in her case to its being so with others. She has the other minds problem.

If she is in fact different from others it makes no difference to her situation. She has as it happens had something left out. But she does not believe this. She thinks she is not different. She thinks others are the same. She moves from what she knows to be true in her own case to its being true for others. Where she is in fact different, she has direct knowledge of the absence of inner episodes. Others have inner episodes and she does not. We might infer that she has no inner episodes. She knows she has not.

In each case her theory of mind holds for her. There are then two possibilities. Either she is hypothetical about others: if they are like me this will hold for them, or she is categorical: this theory holds of others since they are the same. The hypothetical is no use on other minds. The categorical is no use either. How is it known that the theory holds of others, that they are (relevantly) the same?

The following points are crucial. Theories of mind are theories of minds, at least covering all human minds. Our evidence necessarily includes our own case, where we have direct knowledge which we lack in the case of others. A theory of mind cannot be used to establish that there are other minds.

Consider someone trying to make what they can of human behaviour, from the outside. How far can they get? The matter is controversial but the immediate point is that if they are after a theory of mind it has to cover their own case. Assume they insist that they are to be treated from the outside too. That is wilful. It does not assume there is nothing to be learned from the inside, it just sets that aside. It is wilful because there is inevitably something to be learned from their own case. They can learn that there is nothing there at odds with what they are inferring from the outside. They can learn that there is. There are the two possibilities: our own case gives us additional facts; our own case provides nothing. The second possibility is as rewarding as the first. But the rewards depend on a crucial epistemic inference: others are as I am. In the second possibility that becomes: there is nothing additional for others when they go inside.

Consider the Eliminative Materialist. If there are no minds there are no other minds so there is no problem of other minds. That is, however, too quick. Without going into the details of the Eliminativists' position, it is crucial that they insist that the mental is a theoretical category and when all the evidence is in that evidence fails to support the existence of the mental. Here again, one's own case is part of that evidence. There is nothing in one's own "experience" that provides significant counter-evidence to the Eliminativist claim. Nor is it thought that there is reason to think others are different. So the Eliminativist insists that nothing in their own case ultimately supports belief in the mental and further insists that there is nothing in the case of others. But they have direct knowledge only in their own case and these two claims, about themselves and about others, are needed to support their theory. They have to rule out alternative hypotheses about the behaviour of others. This will involve appealing to their own experience in support of their (negative) hypothesis. That there are no other minds depends on the comprehensive claim that there are no minds. But this depends on its being thought that a mind is absent in one's own case and not just theoretically. Evidence on that is crucial, as is the appeal to the similarity between oneself and others. What it is reasonable to believe in one's own case can then be believed to hold of other human figures. One's own case provides crucial evidence both in support of Eliminativism and for setting aside competing hypotheses.

So Eliminativists have the classic other minds problem, transposed to the negative. There are no minds. How is that justified? The asymmetry is once more present. My own case, when examined, and pondered, and finally weighed up, provides no barrier to the negative thesis. But what of others? Perhaps the Eliminativist is different. The old familiar (Behaviourist) tune is heard, perhaps more hauntingly.

It will by now be clear that Functionalism too has the classic other minds problem. If the essence of that view is that all there is to the mental as such is having the relevant role, the Functionalist has to justify just that; there is nothing more to it. Direct knowledge that this is indeed so is obtainable only in one's own case. Perhaps others are different. Direct knowledge is either available or not. If it is not thought to be, again it is only in one's own case that one can report that direct knowledge has nothing to report. That it is the same with others needs to be justified. Functionalism itself cannot be used to show that there are, indeed, others, since it either assumes there are others, who are the same as the Functionalist, or in some way has already supported the view that there are others, who are the same as the Functionalist.

So all theories of mind have the classical problem of other minds. The asymmetry between one's own case and that of others is built in and the theory, being a theory of mind, is a theory of minds (one's own and others). Asymmetry is a consequence of having the capacity for direct knowledge in our own case. Whether or not there is anything there to have direct knowledge of is irrelevant. If the theory denies there are "inner episodes" then it is at least implicit that there is no direct knowledge of such episodes in one's own case. We cannot have direct knowledge that this is so with others. So there is an asymmetry between the knowledge we have of ourselves and the knowledge we have of others. We do not have and cannot have the relevant direct knowledge in the case of others. All theories of mind, therefore, depend on the evidence we have in our own case. They depend on there being nothing from one's own case that is at odds with the theory. Perhaps others are different from ourselves. Perhaps they are the same. All theories of mind need to generalise from one case. They are not epistemological theories so that does not matter. It matters, however, when they are put to epistemic use to show that other minds exist for the theory of mind just assumes that other minds exist and they are the same. It assumes that the theory holds of others in that it assumes that others are the same as the propounder of the theory. It cannot then use itself to show that there are others for a crucial premise will be that the theory holds of others. The theory is not entitled to that premise without epistemological argument (in addition to the other arguments deployed).

Another way of looking at it is this: either there is more to the mental than allowed for by Behaviourism or Functionalism or Eliminative Materialism, or there is not. Let us call this more 'how it feels to the subject' (where this goes beyond awareness that one is disposed to wince or whatever). Let us say that, so understood, qualia exist or they do not. If they do we have, obviously, the (difficult) classical problem of other minds. If they do not exist we still have that problem because only in one's own case is one in a position to know directly that they do or do not. Those who deny qualia, as much as those who affirm them, do so for all of us, not just themselves, and they depend *equally* on the evidence available in their own case. Their "experience" has to support their claim. If the claim is that there are no qualia then the absence of direct counter-evidence in one's own experience has to be accompanied by the *assumption* of its absence in that of others (any argument to that end would be at least partly epistemological). So a theory which has assumed it holds of others cannot then, by assuming that it holds of others, be used to show that there are others.

It will help to compare Dualism and Functionalism's connection with the problem of other minds. Dualism is, of course, not invoked to show that there are other minds because, precisely, if Dualism is true the problem of other minds is a difficult one. But an essential element in the Dualist's position on other minds involves supporting Dualism to the extent that it is shown that it is with others as it is with oneself, so that the required comprehensiveness is obtained. That Dualism causes trouble for other minds is, as we know, seen as a problem for Dualism. The "bottomless pit of scepticism" is thought to loom. But that threat comes from within, from the individual's experience. What we each of us "know" from our experience is at the heart of Dualism, and at the heart of its other minds problem. Correspondingly, the individual's experience has to be at the heart of Functionalism's escape from that "bottomless pit".

Functionalism is used on other minds because it makes the other minds problem easier to solve but it is grotesque to use it in this way. Its comprehensiveness has to be shown whereas the use in question assumes that it is comprehensive. The argument, if it is to be respectable, and work for other minds, needs to go this way: if Functionalism is true, then it is true of others ... The next move will then be: it is reasonable to believe it is true of others ... This will involve the available evidence, where one's own case is crucial for the reasons set out previously. So until we have reason to believe Functionalism is true of other minds we have no reason to believe in its applicability to other minds. In order to believe it is true of other minds we effectively need to have solved the other minds problem. The argument cannot, respectably, go: Functionalism is true, so it is true of other minds, so there are other minds (by, essentially, Armstrong's modest, easy steps). That Functionalism is true, involves its being true of other minds. That it is true of other minds is an essential part of what has to be supported. That it is, indeed, true, needs an argument precisely to that effect.

It we return to Dualism, it is only if Dualism is thought to be true about others that there will be thought to be that particular problem about other minds. We need a reason to believe it is true of others in order to believe that we have the difficult problem. Happily, it follows in this case, that what creates the problem solves the problem. If we know that Dualism is true of others, we must have a justification for that belief, whatever that justification be. Happily, given that I have argued all theories of mind are stuck with the same difficult problem, it is also the case that Functionalism (and all the others) cannot then be less fortunate. It is only if Functionalism is known to be true of others (if it is true at all) that it can be used. Once it is known that either is true of others, as either has to be, if true, we do not need to use it to solve the other minds problem. We neither need to, nor can, legitimately. Dualism and Functionalism sink, or swim, together.

It has been suggested to me (by Keith Campbell) that my argument ignores the complexity that almost inevitably attends arguing for positions in philosophy. At any rate, the particular picture presented is too simplistic. We might argue that a reason for believing Functionalism true is that it would make the other minds problem soluble. One does not then directly argue that therefore Functionalism is true, therefore the problem is soluble. One feeds that reason into whatever others are advanced for believing Functionalism to be true. Together, these reasons constitute the case for Functionalism, jointly persuading one to believe it true. The case for Functionalism is a package. One then puts it to use on the other minds problem.

This seems suspect in that surely the other reasons would by themselves have to be sufficient ground for believing Functionalism true, in this particular context. Here one is directly applying a theory to solve a problem where one of the reasons for adopting the theory would be that it would solve the problem, if it were true. That particular reason should be inoperative, where the theory is to be used to solve that problem.

Otherwise, solving some other problem, it might well be perfectly respectable. In this case, solving this problem would constitute "double-dipping".

However, there is a more specific problem where Functionalism is being used to solve the other minds problem. That it would solve the problem is to be part of the case for Functionalism. But if the rest of the case for Functionalism is hypothetical also then that case is surely weak. The truth of Functionalism would help Materialism so that might be part of the case for Functionalism. Then we would have two hypotheticals. But we need to believe Functionalism <u>is</u> true, not what advantages would apply <u>if</u> it were true.

As I have argued, a reason for believing it true, as opposed to hoping that it is true, would have to be a reason for believing it holds of (at least) human minds, therefore other (human) minds. Such a reason would either have to be transcendental, or would be subject to the argument I have outlined, in that it would have to deal with the asymmetry between ourselves and others. Any transcendental argument to the nature of the human mind would have to be considered on its merits. I do not know of any such argument for Functionalism. The Private Language Argument could be viewed as a transcendental argument. As such, prima facie it could be applied to other minds without offence, but only, of course, if it works, and if it is then relevant to other minds, which is a separate question from its working on its own terms.

The case of Eliminative Materialism is instructive. If it is true there are no minds. But is it true? Here the possibility of a transcendental argument has to be allowed for, but I do not know of one. But such an argument would do the trick, if it worked. Otherwise, our familiar asymmetry applies.

It is even more instructive to consider the use of Functionalism to solve the Mind-Body Problem. If it is true we have solved (by various stages) the problem. That makes it attractive. It is, independently, on its merits, claimed to fit the facts about the mind. So it is believed. So it is then used to solve the Mind-Body Problem in that it is used to defend Materialism against the objection from phenomenal properties (from qualia). This benefit clearly disposes Materialists to embrace Functionalism, but it is claimed, independently of this usefulness, to fit the facts. At the very least, Functionalism has to be seen as plausible, a real option for the mind, on its own account. Its fitting the facts about the mind is an issue, and to be defended, independently of its solving the mind-body problem. That is the crucial difference from the other minds case.

Behaviourism is worth looking at, since there was an argument for it that might (at any rate loosely) be regarded as transcendental. The Verificationist Theory of Meaning, applied to others, resulted in a Behaviourist account of other minds. So Behaviourism, after all, does solve the other minds problem. However, (1) it does so only if the Verification Theory of Meaning is acceptable. (2) That theory itself produces an asymmetry between our own case and that of others, so that for others my pains are constituted by behaviour while for me they are not, *and vice versa*. So that you do not just have asymmetrical first-person and third-person accounts, you have consequently different accounts for the same pains of each person. (3) Behaviourism's/Verificationism's account has to fit our own case, as well as others, since Behaviourism is a theory of minds. Here its "success" with other minds is fatally not repeated. We need a universal account and the Verification theory of meaning fails to deliver such an account. But this third point is what has been argued in this chapter. The point that matters is not that Behaviourism fails that test, but that Behaviourism must undergo it. This detailed account of Behaviourism and other minds emerges from the general pattern. The fact that Behaviourism would solve the other minds problem is "fed into" the case for Behaviourism, which then is adopted, and used to solve the other minds problem. That founders on the fact that either the other arguments for Behaviourism are transcendental, or they must face the asymmetry and argue that Behaviourism covers all minds, our own and others.

It should be noted that any transcendental argument to other minds, or to the nature of mind, or to there being no such thing as mind, would have to be independent of the evidence of one's own case, positive or negative. This would be problematic. How could it be? But one can only wait for the argument.

It has been suggested to me (by Frank Jackson) that we can distinguish between (A) its being the case that everyone must proceed from their own case and (B) its being the case that everyone has the same problem proceeding from their own case. So Functionalists (and Behaviourists) might well respond that they can get from their own case to that of others more easily than can Dualists.

Even if they could, this objection concedes that they do need to proceed from their own case, and they are, therefore, subject to the crucial objection levelled at the traditional analogical inference to other minds, that it is based on one instance, which is claimed to be a hopelessly meagre base (see the relevant discussion in Chapter Four).

But can they, indeed get from their own case to that of others more easily than can Dualists? Any move from one's own case will invoke what is found (or not found) there and will, in claiming that what holds for oneself holds for others, depend on the similarity between oneself and others both in physical make-up and in behaviour. Others are like me will be the cry. As expressed, there seems no room for epistemic differences between Functionalist and Dualist. Each depends on their own case. So depending, whence comes the Functionalist's easier passage to others?

Perhaps it is thought that one can get to Functionalism proceeding from the "outside" and that one cannot get to Dualism this way. But one needs, proceeding from the outside, to rule out Dualism. One needs, independently of that need, to check one's own case. Each (separate) need, ensures a dependence on one's own case which is epistemically crucial. There seems to be no relevant difference between Functionalism and dualism.

It might be thought that if Functionalism is known to be true then our day-to-day, practical access to others is easier than if Dualism is known to be true. I do not find this plausible. We all of us must make do with what we can observe in our dealings with others. Functionalists and Dualists are in the same boat. That there are two distinct theories would seem to leave untouched the practical problems of understanding other lives in specific contexts. But, in any case, that can be cheerfully allowed as a distinct possibility. It does not touch the issues being canvassed here. Such practical access depends on the theoretical access having been negotiated already.

Finally, let me look at the untidy picture outlined earlier in the chapter to see if it can be tidied up even further by seeing why those who went wrong went wrong. What is common ground among the various practitioners is what human beings have in common. They behave in similar ways. They are made of the same stuff. They look much the same. This is observable.

Those traditionally thought to have the serious other minds problem had to show there was something extra (phenomenal features of experience) in all of us, which they claimed to know was there in their own case. Those avoiding (as opposed to evading) the problem, or the serious problem, thought that they did not have to show there was something extra. They knew there was not in their own case, so they thought. Their mistake was to think they did not have to show there was not something extra in the case of others. The traditional problem has been to show the presence of something in others. The error was to think there was no problem in showing its absence in others.

CHAPTER THREE

OTHER MINDS AND SCIENTIFIC INFERENCE*

The fashionable approach to other minds is to view them as theoretical entities. Scientific inference (hypothetic inference, the hypothetico-deductive method) is, consequently, the method used to ground belief in these (unobservable) entities. I have little hard evidence for my claim that this is indeed the fashionable approach to other minds, however. The problem of other minds is not much ventilated in the philosophical literature and when it is, it is unusual to find the suggestion that they are to be regarded as theoretical entities discussed never mind advocated. However, I find that the view is advanced informally, frequently, and it seems clear, anecdotely and otherwise, that the alternatives are generally out of favour. Unsatisfactory as the evidence is, it seems to support best the following options: there is no favoured solution to the problem of other minds; there is generally thought to be no answer currently to the problem; the solution is to treat other minds as theoretical entities. Of these, I am sure other minds as theoretical entities wins hands down.

The consequent method of inference is classically associated with support in science for theoretical (unobserved) entities but it is not limited as a form of inference to supporting such entities. It is the only (inductive) way to support theoretical entities but it has other uses. Given that it is the only way to support theoretical (unobserved) entities it has to be used if minds in general, or other minds in particular, are regarded as such entities. But those who insist that minds are more than theoretical entities can and (as we shall see, do) make use of scientific inference. So Dualists and Reductionists can be found with Eliminativists under this umbrella. Functionalists are there too, no doubt in force, some regarding minds generally (their own and others) as purely theoretical entities, some not, depending on whether they think we have direct knowledge in our own case that minds exist.

The form of this scientific inference (henceforth SI) is this: we are first to come up with a hypothesis to account for certain facts and then we seek to support that hypothesis over alternatives. In the case of other human figures and their behaviour we need to make sense of what we observe. At its most straightforward SI starts with the problem of explaining human behaviour and concludes that the best explanation of that behaviour is that human beings behave as they do because they have minds.[1]

There are various prima facie problems with SI when it is applied to other minds, and particularly when, as clearly with Functionalism, the speciesistic limitation to the minds of human beings is removed.

Firstly, there seems to be a tendency to assume that if we are faced with similar activity then we are to take it we are faced with similar causes. The same explanation is invoked to account for what is observable, provided what is observable is thought to be sufficiently alike. This seems to involve, until it is shown to be otherwise, an implicit dependence on the notion that like effects have like causes, which notoriously wll not do. SI would seem too easily to provide sufficiently versatile computers with minds.

Secondly, lurking behind this dubious invocation seems to be a verificationist bias. It seems to be thought that either a difference that does not manifest itself in some way or other does not exist, or, at least, that there can be no adequate basis provided for the claim that there is, indeed, a difference. The point is that SI is used across the board and, in particular, is to cover human beings and, in addition, non-human persons. It becomes either too easy to proliferate minds (beyond necessity) or SI has to be supplemented by some other reasoning, so that this too easy proliferation can be reined in (the analogical inference clearly beckons ...).

Thirdly, although it is important that SI can be, and is, used indifferently by those holding Dualist, Functionalist or other theories of mind, there is a tension between what one would expect of this application (of SI) and what is (almost) invariably the case. The tension is between the expectation that those viewing other minds as theoretical entities would treat minds in general (including their own) that way, and the fact that almost invariably it is only other minds that are treated that way. In the case of Dualists, say, the tension is between the rich content of their own minds as experienced and their treatment of other minds as theoretical entities (how can we fill in, in this wise, so much "other" content). In the case say, of Functionalists, there is the contrast between the "impoverished" content of minds in general and the fact that their own minds are available to introspection. There does not seem much to introspect.[2] But if there is, the richness of what is introspected seems to hang loose, to dangle from SI.

These remarks are, however, preliminary and informal. They can be allowed to stand or fall as they show up in my considered case against SI. That case is that SI is unable to support belief in other minds without appealing to one's own case. The appeal is epistemically crucial. SI is, therefore, not available as a non-analogical alternative to the analogical inference to other minds, supposedly free of that notorious inference's crucial weakness. However, SI's attractiveness to its adherents is largely that it is, indeed thought to be such an alternative. It is indeed, embraced precisely because it is thought to avoid the crucial problems that beset that particular analogical inference. The two classical objections to the analogical inference to other minds are that it has a logically uncheckable conclusion and is a generalisation from only one instance. SI does not avoid the former objection, being itself an inductive argument with a logically uncheckable conclusion. That feature, however, is no longer seen generally as problematic. My argument will show that SI does not avoid the latter objection (generally seen as fatal) since it cannot support its conclusion without crucially appealing to one's own case. SI is as weak in this respect as the analogical inference to other minds.

The argument will proceed in stages. First I show that Dualists cannot use SI without appealing to their own case. I then do the same for Reductionist Materialists. The argument is then generalised by looking at Functionalism's use of SI. All cases of SI will be seen to require an appeal to one's own case which is epistemically crucial.

The key move in the analogical inference to other minds, in support of the conclusion that other human figures have minds, is from what each of us knows to be true in our own case, to how it is with others. That inference appeals to the similarity between ourselves and others to support the belief that others are as we are in having minds. Clearly, this is to argue by analogy, from what is known to be true in one case to its being true in other (like) cases.

I have characterised SI, in its simplest form, which is also, I think, its standard form, as concluding that the best explanation for other human figures behaving as they do is that they behave as they do because they have minds. In this form SI depends on the claim that mental states cause behaviour, so setting itself against Epiphenomenalism. Whether or not Epiphenomenalism is true or false is not the issue here. What is relevant to SI, and, ultimately, fatal to SI, is the Epiphenomenalist belief that human behaviour is caused by physical (brain) states not by mental states. The Epiphenomenalist believes that the behaviour of other human figures is caused by physical (brain) states not by mental states. So Epiphenomenalists cannot use this form of SI.

Notice, however, that the dispute between Epiphenomenalists and other Dualists is whether mental states should be assigned a causal role. They agree that they exist. Epiphenomenalism is a sort of half-way house for Materialism. According to Epiphenomenalism what happens in the universe has purely physical causes, though mental events occur. The dispute between them and Interactionists does not turn on whether purely physical causes are in principle to be ruled out. The dispute turns on whether, given that mental events occur, they have or have not a causal role. In particular, those Dualists (Interactionists) who believe mental events make a difference, have a causal role, depend for their response to Epiphenomenalists on the fact that mental events exist. Their argument is not that purely physical causes could not account for human behaviour. The various arguments against Epiphenomenalism all depend on the existence of mental states and some anomaly that is claimed to follow from regarding them as causally inert (most famously, perhaps, the wild implausibility of the belief that minds make no difference).

What follows from this dependence is that Dualists, faced only with human behaviour, and ignoring their own case, should conclude that human behaviour is caused by purely physical states, in particular, brain states. Epiphenomenalists will, of course, reach that conclusion. But so should Interactionists, since they will not, within those limits, be able to appeal to the existence of mental states to which they can then assign a causal role. They will be forced in such circumstances to conclude that the behaviour of other human figures has a purely physical cause. What blocks this is their belief that mental states exist in their own case and have a causal role in that case. Unless blocked in this or some other way, the inference to purely physical causes is the most reasonable one.

The same problem exists for Reductionist Materialists (my terminology contrasts Reductionist Materialists with Central-State Materialists and Functionalist Materialists). Reductionist Materialists think mental states exist and exist as traditionally conceived, but are contingently identical with brain states. However, as argued in Chapter Two, they have to show that mental states exist in the case of others. If mental states are brain states then as such they will be observable but they will not as such be observably mental states. The problem for Reductionist Materialists if they seek to use SI to show that mental states other than their own exist can be presented in the form of the threat posed by Eliminative Materialism. (It could equally well be presented through Epiphenomenalism. The latter thinks mental events exist and the former thinks they do not, but the difference is irrelevant to the damage each causes to an attempted Reductionist use of SI.)

Eliminative Materialism believes there are no mental states, tout court, whether traditional or otherwise. So it rejects the claim that the behaviour of others is caused by mental states. Whether Eliminative Materialism is true or false is, once again, irrelevant here. If our evidence were limited to human behaviour Eliminativism would not be the minority taste it is. What would there be to eliminate? The Reductionists disagree with the Eliminativists because they think that mental states exist and that they are not purely theoretical. They are experienced (at least some of them are, e.g. pains). Reductionists believe that mental states exist as traditionally conceived, complete with qualia. In the case of others, the Reductionists have first to show that such mental states exist in the case of others in order for them to support the claim that they are responsible for the behaviour of others. SI cannot give that result without the crucial appeal to their own case. This is needed in order to block the alternative, Eliminativist account.

It is helpful to consider the particular case of pain. Reductionist Materialists agree with Dualists that pains exist, as traditionally conceived (at any rate, they agree provided no ontological baggage is carried). There is no fundamental disagreement between them about the concept of pain. The disagreement is over whether or not pains are purely material. That disagreement does not turn in this case on differences about the concept of pain.

Do other people have pains? Eliminativism shows that SI, by itself, cannot provide Reductionist Materialists with other pains. It will give them purely neurophysiological (brain) states lacking the phenomenal properties of pain. How would they reach pain (as traditionally understood) merely through the behaviour of other human figures? Setting the Eliminativist hypothesis aside requires that we use our own case.[3]

Take the traditional case of wincing. Looking from the outside we can extract an internal state responsible for the wincing but we cannot extract pain as traditionally understood. The two competing causal hypotheses, the Eliminativist neurophysiological internal state, unconnected with traditional pain, and traditional pain, have to be assessed respectively as the cause of wincing. Only the introduction of our own case would enable us to give Reductionism the nod over Eliminativism.

So consideration of Epiphenomenalism and Eliminitavist Materialism shows that neither Dualists nor Reductionist Materialists can use SI to support their belief in other minds where such minds are understood traditionally. SI needs to be supplemented by an analogical appeal to one's own case in order to set aside the competing hypothesis, that the behaviour of other human figures has a purely neurophysiological cause, which lacks the traditional phenomenal properties of mental states.

None of this seems to create any problem for Central State Materialists and (strict) Functionalists. Both believe in physical (material) causes and neither accepts the mental as traditionally understood.[4] So it would seem they can use SI to support their beliefs in (their) other minds. Take the Functionalist Materialists. They believe in the mental but believe the realising states are physical. Unlike the Reductionists they have a non-traditional account of the mental. For them, to be a mental state is to be an internal state with a particular role involving inputs and outputs (and other internal states but that can be ignored in this context). What makes a state a mental state is purely a matter of the role it plays.

There would seem to be nothing to prevent a Functionalist using SI successfully, faced only with human behaviour. Given the inputs and the outputs we plug in the relevant mental state. If pain is whatever is (typically) caused in certain circumstances and (typically) causes certain behaviour then faced with those circumstances and that behaviour we hypothesise that pain is present (we slip pain into the slot). That the particular pain will as a matter of contingent fact be a neurophysiological (brain) state will not prevent its being pain if it occupies the relevant role between the circumstances (input) and the behaviour (output). SI and Functionalism seem made for each other.

However, the lesson of Chapter Two is that this cannot be so. Chapter Two has taught us that no theory of mind avoids the classical problem of other minds. So Functionalism despite perhaps being partly fashioned to avoid that problem will not have succeeded. Being a theory of mind it has to hold generally of all minds so it cannot be used to show there are other minds. That is its problem in a nutshell. As a theory of mind, it was shown in Chapter Two that it needs to appeal to the facts available in the propounder's case. The epistemological asymmetry between one's own case and that of others applies to all theories of mind and that is why each theory of mind needs to appeal to its propounder's case. SI is, of course, supposed to avoid what the argument of Chapter Two has shown to be unavoidable. So it follows from Chapter Two that SI is not available for any theory of mind, whatever that theory be, as an alternative which avoids using an inference based on one's own case, since that chapter shows that all theories of mind have the same other minds problem and all need, as theories of mind, to appeal crucially to the facts revealed in their propounder's case.

So in advance we know that SI and Functionalism will turn out not to have been made for each other. We know in advance that SI will not work for any theory of mind, will not avoid the crucial problem. Let us see how it fails.

SI comes a cropper in this way. Any hypothetic inference prefers a particular hypothesis to others. There are always alternative hypotheses. They have to be set aside in favour of the preferred hypothesis. For Dualists and Reductionist Materialists the problem is simple: they are unable, using SI, to set aside the particular competing hypothesis that other human figures do what they do because of purely physical, non-mental factors. To have any chance of doing so, they need their own case.

However, any theory of mind has to set aside the competing hypotheses that are at odds with what it believes about minds (including minds). Standing the problem for Dualist and Reductionist Materialists on its head, Functionalists have to set aside the competing hypothesis that the behaviour of other human figures is caused by mental states as traditionally understood (with features additional to those allowed by Functionalism). What the Functionalist believes about other minds is distinct from and incompatible with, what the Dualist and Reductionist Materialist believes about other minds. The latter belief is not just a possible hypothesis. It is a belief held widely. It is a working alternative. It cannot be ignored. It needs to be set aside.

The setting aside is a matter of supporting Functionalism's view of the mental against the traditional view of the mental. Supporting the Functionalist view of the mind involves various matters, but one is that there is nothing in one's own case that creates a problem for the Functionalist view. An element in the Functionalist case has to be the invocation of one's own experience. One cannot invoke that of others. One's direct knowledge is limited to one's own case, even if that direct knowledge is negative, reporting the absence of any features in experience that would work against the Functionalist view.

This is, of course, to rehearse the argument of Chapter Two. But in the present context it is to be noted that attacking the traditional analogical inference, as a generalisation from a solitary instance, will be too comprehensively successful. The problem posed by the excessively, outrageously meagre inductive base is shared by all theories of mind. The traditional view of the mental, if one is a Functionalist, has to be set aside in one's own case. But it is only in that case that one can move so directly against it. The generalisation from that case is then a common burden. One cannot get at other minds (Functionalist or otherwise) directly. So the argument of Chapter Two showed that SI had to fail for everyone, for that argument showed that theories of mind cannot legitimately be used to solve the other minds problem (or even make it easier). They each, as theories of mind, have to invoke the propounder's experience. All theories of mind are in the same epistemological boat in the case of other minds. The way SI, if followed through, succumbs, as it must, is in the need to set aside competing hypothesies. That is the point of fracture. that is where we find the inevitable appeal to one's own case.

So, since everyone has the classic problem of other minds SI does not work by itself. Everyone has recourse to their own case. How then is what is learned from what one's own case used to show there are other minds? I shall begin with the Dualist situation.

It is widely held that insofar as we have direct knowledge of our mental states, we do not have direct knowledge that they have a causal role. Whether or not they have such a role is a theoretical matter, given that they exist and bear certain relationships to our behaviour. On this view it is not that we believe, from an inspection of our mental states, that our behaviour is caused by them. Rather, we know our mental states exist and are correlated with our behaviour; we hypothesise that those states have a causal role. (Of course, if we are Epiphenomenalists we hypothesise that they do not have a causal role.) So the hypothesis is not: my behaviour is caused by mental states. The hypothesis is: my behaviour is caused by the mental states which I know (directly) exist. The crucial knowledge is that my mental states exist. The causal hypothesis assigns to those mental states a causal role. Concerning the behaviour of others that hypothesis cannot tell the desired causal story until the existence of the relevant mental states is established - to which it can then assign a causal role. So we cannot move directly from my causal story to that story's applying in the case of others. My causal story depends on the belief that my mental states exist. The causal story for others will depend on the belief that there are, indeed, mental states belonging to others. The existence of the mental states come first.

The attempt to get directly from what causes my behaviour to what causes the behaviour of others falls foul of the fact that if our evidence is limited to facts about others then the most likely explanation of human behaviour is that it has a purely physical (non-mental) cause. The inference that what is true in my case will be true in the case of other human figures, depends on there not being another, preferable, explanation. But there is; that their behaviour has a purely physical (non-mental) cause. What prevents our believing this in our own case is, firstly, our direct knowledge that our mental states exist and, secondly, our opting for the hypothesis that these mental states have a causal role. Unless blocked in the same way in the case of other human figures, the preferred explanation will be that physical causes account for their behaviour. That explanation, if it is to be blocked, is blocked by inserting the premise that other human figures have mental states. It cannot otherwise be blocked.

Can this premise be supported? It can by these steps: I know that my mental states exist; I hypothesise that they are caused by some of my physical states; I note the analogy existing between my physical states and the physical states of others; given a belief in the uniformity of nature that like causes have like effects, I infer that the physical states of others will cause mental states in others. Only then, if my hypothesis about mental states is that they are causal, can I apply this to infer that the behaviour of other human figures is caused by the mental states of other human figures. If I am Epiphenomenalist I will not do so, of course, and, more importantly, I need not.

It might be claimed that supporters of the view that their behaviour is caused by their mental states can infer that similar behaviour in similar organisms will be caused by mental states. It might be claimed, further, that such a view is as respectable as the inference I have just advocated. But this suggestion depends on the highly dubious principle that like effects have like causes. This objection is sometimes responded to, perhaps in the spirit of Mill, with the claim that the inference is reasonable in the absence of plausible alternative explanations. This response, however, is unacceptable. For there <u>is</u> a plausible alternative explanation, viz., that the behaviour of others is caused by non-mental, purely physical states. The only reply that seems available is to cite one's own case as counter-evidence. But this can only be effective if it is reasonable to infer that like effects have like causes. Which takes us back to where this line of thought started.

So this alternative inference fails. Given its failure, direct knowledge of the causal role of our mental states would not affect the form of a viable inference to other minds since such direct knowledge, if it were to make a difference, would have to issue in such an inference (from what causes my behaviour to what causes the behaviour of others).

The outline above of how Dualists reach other minds also holds for Reductionist Materialists. They have a metaphysical and ontological difference of opinion with Dualists. They believe that mental states are, as a matter of fact, one and the same as brain states and have only material properties. This, however, makes no epistemological difference. Reductionist Materialists believe mental states are brain states. As outlined earlier, in the case of other human figures, Reductionist Materialists first have to show that there are, indeed, mental states to be identified with brain states. The observability of brain states is no help since they are not observably mental states.

Does Functionalism make any epistemological difference? They have the classical problem of other minds. SI will not work for them without crucially appealing to their own case so that they will be enabled to set aside competing hypotheses. So it is unlikely that there will turn out to be for them a different route to other minds. Their view of the mental depends crucially on their own (phenomenal) experience (even if that be construed as the absence of phenomenal experience). Functionalism is in accordance with that experience. Their crucial inference is that others are as they are. Others have minds, internal occupiers of roles without other properties, just as they do. Others are not different. Whether or not the Functionalist claims to have direct knowledge in their own case that such internal states exist is irrelevant to the fact that their own experience is crucial. Nothing in that experience causes trouble for them. The inference that others are the same is dependent on their similarity in make-up and behaviour. That is involved in supporting the Functionalist belief about other human beings. What is believed to hold for oneself is thought to hold for others. That involves setting aside the Dualist belief about other minds. That depends on appeal to one's own case. That there is nothing in that case that blocks the Functionalist hypothesis is crucial.

The same epistemological story holds for Eliminativists. They have to support their view of the mental by appealing to their own experience. They have to rule out alternative hypotheses about the behaviour of others. This too will involve appealing to their own experience. There are no minds so there are no other minds. But this depends on its being thought that a mind is absent in one's own case. Evidence on that is crucial, as is the appeal to the similarity between oneself and others. What it is reasonable to believe in one's own case can then be believed to hold of other human figures. One's own case provides crucial evidence both in support of Eliminativism and for setting aside competing hypotheses.

So the various theories of mind reach other minds (or their absence) in the same way.[5] This is as one would expect since they all have the same problem of other minds and, as Chapter Two showed, a theory of mind cannot itself be used to solve the problem of other minds. It is also what one would expect given that theories of mind are theories of minds. They have general coverage. Each of us is covered. Each of us is the same. Link this with the epistemological asymmetry between each of us and others and we get the crucial importance of the evidence available in our own case. To get from that to others we need the similarity between ourselves and others. So the variety of theories of mind does not issue in a variety of solutions to the other minds problem. This is as one should expect. Theories of mind are metaphysical theories. The other minds problem is an epistemological problem.

CHAPTER FOUR

THE ANALOGICAL INFERENCE TO OTHER MINDS*

If we all have the problem of other minds, if no theory of mind manages either to avoid the problem or provide a solution, if there is no way of avoiding crucial use of one's own case, and if a scientific inference fails, on its own, to justify our belief in other minds, then it is to be expected there will be renewed interest in the (hybrid) analogical inference to other minds. We often make do, in philosophy at any rate, with what we have, and the absence of an alternative is, in philosophy, generally fatally seductive. For many philosophers, scientific inference was embraced to deliver them, if not from the bottomless pit of scepticism, at any rate from the analogical inference to other minds.

This chapter is intended to reassure them. It will defend the analogical inference to other minds against those objections that claim it is a bad inductive argument. Nothing in this chapter turns on the fact that the analogical inference being defended is a hybrid of the traditional analogical inference and a scientific inference. Other chapters will respond to claims that it involves, in some way, conceptual incoherence or confusion.

There are two classical objections to the analogical inference to other minds as an inductive argument. One is that it is *logically impossible* to check up on its conclusion. The other is that it is a generalisation from one case (often poignantly referred to as a solitary instance). If the two objections are put together, then the argument will be a generalisation from one case, as a matter of logical necessity. However, that the generalisation's meagre base is as meagre as it is of necessity does not seem to matter to opponents of the analogical inference. Nor should it. I shall begin my defence of that inference by considering the objection that the inference is a wild generalisation from one case.

The standard reply to this objection is to insist, in effect, that the generalisation is not understood, if it is regarded as a generalisation from one case. The reply is most famously associated with A.J. Ayer. I shall argue that this reply fails.

If "one case" means that the evidential base is observation of just one occasion of a mental state being correlated with a physical state, then the objection would be simply vitiated by a false premise. It is obviously not true that the analogical inference to other minds is based on one case in this sense. It is based on the multifarious correlations between mental and physical that one has observed in one's own case. On the other hand, so the standard reply goes, if "one case" just refers to the fact that these correlations between mental and physical are all drawn from one's own case, then the objection is merely drawing attention to the fact that the members of the sample on which the analogical inference is based have a common property; namely, that of being correlations in one's own case. And it is quite generally the case that analogical arguments proceed from samples having common properties. For example, helium, like all the elements, has a characteristic spectrum. The nature of this spectrum is determined by experiments carried out, of course, on earth. It was then observed (actually, remembered) that light from the sun displayed this spectrum (along with hundreds of others), and from this it was inferred that helium was present on the sun. In this argument, all the samples of helium were on earth.

This standard reply is, however, unacceptable. It is true that there is a veritable multitude of correlations in one's own case between mental and physical goings on. However, the analogical inference to other minds is, in fact, when properly understood, an analogical argument from only one case, rather than an analogical argument from a number of cases having a common property. Two different considerations, taken together, show that the analogical inference to other minds is rightly regarded as an argument from one case.

The first consideration turns on the point that it is just false that every, rather than most or many, instances of, for example, pain behaviour on one's own part are accompanied by a pain of one's own. This means that if we regarded the analogical inference as proceeding from the many instances of mental-physical correlation that hold in one's own case, the only conclusion that could be obtained would be: many instances of behavioural states are associated with mental states. Now this conclusion is just not the conclusion we need, because it is consistent with certain human bodies which behave just as one's own does, not having *any* associated mental states.[1] The conclusion we want is that for every body which behaves much as one's own does, it is true that many instances of its behavioral states are associated with mental states. So the standard reply fails.

The second consideration arises from the fact that we believe that for each living human body there is at least one person such that the behavioral states of that body are accompanied by the mental states of that person.[2] As we shall see, even if the standard reply succeeded, its conclusion, that other human bodies have mental states associated with them, is unable to sustain the claim that there is at least one person associated with each body. The only analogical argument that could show this would be one generalising from the fact that for one's own body there is at least one person (oneself) such that the behavioral states of that body are accompanied by the mental states of that person. Now this is, and must be, an inductive generalisation from one case; therefore, if the analogical argument for other minds is to establish that for each living body there is at least one person whose mental states correspond to the behaviour of that body, the argument must be regarded as an analogical argument proceeding from one case.

However, R.I. Sikora[3] thinks the standard reply is 'basically correct'. But Sikora claims that 'when properly understood, the argument from analogy is not in itself an argument for other minds'. In his view, other minds are reached in two stages. The first stage, which is truly an anological inference, can indeed show that 'experiences are associated with other bodies' and thus meet the minimal requirements that, for each human body, there are associated experiences. At this stage one appeals merely to the 'enormous number of correlations between instances of behaviour and experiences'. This stage is a multiple-case analogical argument. One then shows, using hypothetic inference, that 'at least one person is associated with every other body during its active existence'. This second stage is not an analogical inference never mind a single case analogical inference.

Sikora's response is worth close attention since it is the only attempt I know of to circumvent the unpalatable insistence that the analogical inference to other minds has to be a generalisation from one case. I shall argue: (i) that no multiple case analogical inference will support the conclusion that experiences are associated with every other body, and (ii) as one would be more than prepared for by now, that even if it did, the second stage hypothesis if it is to support the desired conclusion, needs evidence obtainable only in one's own case, and so fails to escape the problem of the solitary instance.

I have claimed that any multiple-case version of the analogical inference to other minds would fail to establish that every other human body had associated mental states. Here is the supporting argument for my claim as formulated by Sikora (on his way, he hopes, to refuting that claim):

> (i) since I only observe that most of my manifestations of pain-behavior are accompanied by pain, I can only infer that (ii) many instances of behavioural states are associated with mental states and that it does not follow from this that (iii) the common sense belief that *every* body which behaves in a manner similar to my own will have many instances of mental states associated with its behavioural states is correct. (p. 139)

Sikora's response seems to accept that (iii) does not follow from (ii). His tactic is to claim that:

> On those occasions when I engage in pain behaviour but am free from pain, my pain behaviour is nonetheless associated with *some* other mental state. Thus one is entitled to infer that, for *every* body which behaves much as one's own, that many instances of its behavioural states are associated with mental states. (p. 139)

Unfortunately, this tactic is not generally available. Presumably Sikora has in mind pretence. But we all behave to some extent when asleep, and some sleepers engage in behaviour of some complexity. So it is not the case that we are aware concerning every relevant piece of behaviour, that it is accompanied by a mental state. Also, there is the phenomenon of 'coming to' when we have been behaving in a certain way, say driving a car, and in such cases we may have had no awareness of any accompanying mental states. Such cases in particular are thought to be responsible for drivers losing control, when the hypothesis is that they have gone into a trance/coma, and for some time before they crash. This is only a dramatic instance of a quite common general lack of awareness of one's behaviour. A person under stress, or concentrating on some task, may not notice sufficiently how they are behaving in order to provide Sikora's tight correlations. In such cases it is not clear that they have been continuously conscious. So we are just not aware of every relevant piece of behaviour that it is accompanied by a mental state.

Accepting that we believe that all other human bodies are associated at some time with specific sorts of mental events, Sikora moves on to consider the argument modified to deal only with pains:

1. Usually I feel pain when I exhibit pain-behaviour.

2. Therefore many instances of pain behaviour are associated with pain.

3. It is nonetheless possible that when certain human bodies exhibit pain behaviour, this pain behaviour is never accompanied with pain.

4. Conclusion: A multiple-case argument from analogy fails to show that it is certain that all human bodies are sometimes accompanied with pain. (p. 139)

It is the second premise that is crucial: many instances of pain behaviour are associated with pain, but not all.

Sikora accepts this argument but claims:

> its conclusion isn't particularly damaging to multiple-case arguments from analogy. It can be admitted that we can't be certain that for every single human body some instances of its pain-behaviour are associated with pain, so long as the possibility is sufficiently slight. It might be objected that the plain man wouldn't admit that there is even an extremely slight possibility of such exceptions. However, a philosopher needn't be unduly disturbed if this turns out to be another case where we can only hope for moral certainty, even if the plain man thinks (or is supposed to think) that we can have absolute certainty. (pp. 139-140)

This will not do. The trouble is not that we are left to speculate about what might be meant by 'moral certainty' and 'absolute certainty'. The fatal bite of the argument is in what Sikora concedes, that what can be established is only that most (even if an overwhelming majority of) instances of pain behaviour are accompanied by pain. Given an unequal distribution, this could be true even though several bodies one is frequently confronted with behaved as if they were in pain but, in fact, never were. The observational base does *not* establish that it is sufficiently certain (though not absolutely certain) that *every* body that behaves appropriately is, at least sometimes, accompanied by pain. Without further evidence, we are in no way justified in claiming that *every* such body is so accompanied. All the observational base can support is that some such bodies are so accompanied (enough to supply the required measure of instances of pain-behaviour accompanied by pain). We cannot infer from this that it is even likely that all such bodies are so accompanied (nor, of course, that it is unlikely). We are not entitled to an opinion one way or the other. So, given the evidence, we are not entitled to claim that 'there is only an extremely slight possibility of such exceptions'.

Nor are we entitled to our belief of any particular body that it is accompanied at least sometimes by pain. Even if we were sure that most human bodies are so accompanied, we are not thereby assured of any particular such body that it is so accompanied. In the absence of further information, we are not in a position to say whether there are exceptions and, if there are, which they might be. The evidence we have is neutral on whether or not it is likely that the particular body we are confronted with is an exception. Assurance on this (moral or otherwise) is crucial for each of us.

The point may be illustrated this way: take an outbreak of head lice among schoolchildren in a district. Perhaps it is known that 30% will suffer. Not only does it not follow that for each school 30% will suffer, it is known this will not be so. Overall 30% will suffer but the distribution is uneven. Likewise, from the fact that say 80% of behavioural states are accompanied by a mental state it does not follow that for each human body 80% of its behavioural states are accompanied by mental states, nor is this even probable given the evidence. What is true is that the *average* will be 80%. And without further evidence we can say nothing about the probability or improbability of *wide* divergences from this figure in the case of some bodies.

So we are not entitled to be even 'morally' certain that for every human body some instances of its behavioural states are associated with mental states. Further, it should be understood that the range of inference is over all human beings that ever have and ever will exist. If there are exceptions, no matter how few, perhaps it is my personal misfortune that some of my closest friends are contributing the exceptions, or my suburb is contributing them, or this period of history is contributing them all. Perhaps the thought that aliens might have taken over the bodies of those around me would be less upsetting (at least I would not be alone). Again, it is not that there is only a slight possibility of there being such exceptions. On the only evidence Sikora has available, there is no reason to believe there are or there are not.

Sikora in fact indicates later that he is aware of this kind of difficulty for he next considers a revision of the modified argument:

> I observe a number of correlations between my own pain and my own pain behaviour. However, I also note that there are a small but significant number of occasions when I exhibit pain behaviour but feel no pain. Accordingly, whenever I observe pain behaviour in another person, an analogy with myself would lead me to conclude that there is a small but significant chance that he is not in pain. It follows from this that if I were to observe all the pain behaviour of everyone who exists, statistical considerations would make it fairly probable that there are some human bodies such that no instance of their pain behaviour is accompanied by pain. Therefore I am not entitled by a multiple-case argument from analogy to be as sure as I am in fact entitled to be that every other body is at some time associated with pain. (p. 140)

His response is to claim that a multiple case analogical inference 'can be buttressed with too much additional evidence for the conclusion to survive'.

In this passage Sikora is considering a position notably stronger than mine, in that the suggested inference is that 'it is fairly probable that there are some human bodies such that no instance of their pain behaviour is accompanied by pain'. The position I am arguing for is that on the evidence cited we are not in a position to have an opinion one way or the other. Both positions, of course, would be equally unsettling in practice. Since Sikora's replies to the stronger claim could be as well directed to mine I shall consider them as if they were, indeed, so directed.

The buttressing operation referred to proceeds by whittling away at the exceptions to the required physical-mental correlations. The first buttress depends on the claim that 'there are only certain circumstances in which I exhibit pain-behaviour when I am not in pain'. Given this, observation over a long period would eliminate them. But, as has been frequently pointed out, perhaps most famously by Norman Malcolm,[4] the list of such circumstances is indefinite. Given this, it is not possible to note them, then note their absence, and thus conclude another body is in pain.

The second, and 'more important' buttress, involves adding correlations between pain stimuli and pains,

> because in the case of certain kinds of pain stimuli, there is a far greater correlation between my being subjected to the stimuli and feeling pain than there is between my exhibiting pain behaviour and feeling pain. (p. 140)

Unfortunately, this response is not generally available, since there are many mental states not discriminably correlated with stimuli. More specifically, we are not told which kinds of pain stimuli are invoked. Insofar as Sikora has in mind stimuli which cause severe pain, then many human beings will remain in question since they are, mercifully, never exposed to them. Insofar as he moves away from such stimuli, his claim that the correlation is far greater becomes more and more dubious (notoriously, pain thresholds differ, while various circumstances both internal and external change how external stimuli affect us).

Sikora's final buttress, providing the final tightening which will bring security,

> is the invocation of correlations between neural states and feelings of pain which would presumably be even tighter than the correlations between pains and pain-stimuli. Any doubts that remained after this would I think be perfectly tolerable. (p. 140)

One might think that the mildest of doubts (the mildest of epistemic "pain") would, if continuous erode one's contentment, but, in any case, it seems that current neurophysiology would reject this claim. The same sort of mental event may be correlated with different kinds of neural states, even in the same individual. There seems not to be the one to one correspondence Sikora assumes (and needs).[5] In any case, such a sophisticated invocation would be available in practice to very few (until recently to none) and unavailable even in principle to a large number of human beings.

So the extended buttressing is unacceptable. In any case, even were it to be acceptable it would not meet the thrust of the attack. If the relevant correlation has exceptions then we are thereby unable to infer that every human body that behaves appropriately is, at least sometimes, accompanied by appropriate mental states (or, indeed, accompanied by any mental state). Sikora's buttressing is not even in intention designed to produce or thought to produce the required exceptionless correlations.

I have argued that his multiple case analogical inference will not sustain the conclusion Sikora needs. So the first stage of his two-tiered approach fails. I will now show that even if this were not so, the use Sikora makes of it in his second stage will not succeed without an appeal to one case, one's own. It was made clear, in Chapter Two, that this will be so. However, Sikora's second stage fails in a particular way which nicely complements and corroborates the more general argument.

Sikora's idea is to establish first, with respect to other bodies, using the multiple-case analogical inference:

> that (i) those bodies have been associated with mental events throughout their lives, and (ii) that some of those mental events are ostensible memories that resemble my own ... (p. 137)

Once this is done he claims he can

> justify the hypothesis that at least one person is associated with every other body during its active existence (hypothesis PB) in much the same way as I can justify the similar hypothesis in regard to myself. (p. 137)

How is this effected with regard to oneself? Here is how it is done:

> At any given time, I have many memories of experiences which I seem to have had in the past. None of these memories associate these experiences with other bodies, and though some memories may not be associated with any body, I have other memories which make it reasonable for me to suppose that all of my experiences are causally dependent on my body in a way in which they are not dependent on the body of anyone else. I have had, of course, many more experiences than I can remember but there is no reason to suppose that the experiences I cannot remember differ from the ones that I can in a way that would justify me in thinking that the former were associated with some other body rather than with my own. On the basis of these considerations it is reasonable for me to infer that all of my experiences have been associated with the same body, and therefore that at least one person has animated my body throughout its existence. (p. 137)

Essentially, the move in one's own case is to claim that one's experiences are such that PB is the preferred hypothesis. In the case of other bodies, the claim is that the experiences associated with those bodies are such that PB is also the preferred hypothesis:

> When I turn to other bodies, I need only the suppositions (obtained, as I shall argue, through analogical arguments from a multiplicity of cases) that (i) those bodies have been associated with mental events throughout their lives, and (ii) that some of those mental events are ostensible memories that resemble my own, in order to defend a hypothesis for other bodies similar to that which I formed about myself. (p. 137)

But it is crucial in one's own case that one has had no experiences associated with other bodies. This is supported by the absence of memories associating any experiences with other bodies. So it is not the case that the two suppositions cited are all that are needed. That some memories fit the required pattern is consistent with others' failing to do so.

How is the crucial evidence to be obtained for another body (without appeal to the outlawed single case analogical inference)? The only way would be to infer from the absence of the relevant memory reports that the relevant experiences were absent. But in the absence of such experiences there can be no direct evidence that such experiences would be reported. It is clearly false that all memories are reported (most are not). So recourse would have to be had to the highly dubious claim that for any sort of experience at least some memory reports would be made. This is doubtful enough if interpreted as holding that some bodies at least on some occasions would issue relevant reports. It is surely false if claimed to hold for each of us. It would have to hold for each of us since that is the only observational base available in this context. The mere claim that such an experience would surely be reported is unconvincing in the absence of such an experience. I conclude that PB will not work for other bodies.

Finally, to return to the theme of my earlier criticism of this multiple case analogical inference, it should be noted that memories play the crucial role in Sikora's stage two hypothetic inference. The correlation between memory reports and memories is far from exceptionless. Further, the buttressing appealed to in defence of his multiple case analogical inference stage 1 would seem unavailable for tightening this particular correlation. So it is not clear why even Sikora would think he can show, in stage one, that for every human body there are associated memories.[6]

It seems clear then that the analogical inference to other minds is, ineluctably, a generalisation from just one case. I think that is what each of us intuitively knows (and what philosophers fear). If the analogical inference is to be embraced, that generalisation has to be embraced also. Just that is the major objection to be considered below. However, I shall consider first an objection sometimes confused with that objection. This is that the ratio of examined to unexamined instances is not high enough. Don Locke complains that the analogical inference to other minds

> proceeds from the smallest possible basis to the largest possible conclusion; it argues from one solitary instance to, perhaps, an infinity of cases.[7]

This objection involves a misunderstanding about the nature of the analogical inference to other minds, in that it involves treating that inference as a species of *statistical* argument. In fact, as the following argument shows, the soundness of the analogical inference to other minds is independent of the ratio of examined to unexamined cases.

Suppose I use the analogical inference to other minds in an attempt to establish that just one other person has a mind, then no complaint can be made about the ratio of examined to unexamined cases because this ratio will be very much higher than in the vast majority of cases of inductive generalisation. But, clearly, this one other might be *any* other. Therefore, if it is reasonable to use the analogical inference to other minds to show that one other person has a mind, then, for any other person, whoever that person be, it is reasonable to use that inference to show that they have a mind; provided, of course, that they are behaviorally, and otherwise, sufficiently like myself. The reason why the ratio of examined to unexamined cases is irrelevant to the soundness of an analogical argument is that the soundness of an analogical argument depends on the relevant analogy between the examined and the unexamined cases. If the analogy between the examined cases and some given unexamined case is sufficient to sustain the inference, then, by parity of reasoning, in *any* unexamined case where the analogy holds, regardless of how many of these there are, we are entitled to make the inference in question.

The objection that matters to the analogical inference is that put by, among others, Norman Malcolm. It is this objection that is almost universally thought to be successful, and fatal. It does look to be a problem which demands a satisfactory solution. Malcolm says

> ... this would be very *weak* inductive reasoning, based as it is on the observation of a single instance.[8]

This objection turns, not on the ratio of examined cases to unexamined cases being low, but on the fact that there is only one examined case. As such, it is *prima facie* a good objection because it is indisputable that we often require arguments from analogy to be based on more than one instance.

However, this objection should not be confused with the objection:

> How do I know that it is not my individual peculiarity to feel angry when I frown?[9]

One may enquire of any inductive base, how do I know that it is not a peculiarity of the sampled cases to be as they are? Since it can be raised against *any* analogical argument it is not an objection to the analogical inference to other minds as such. This holds also for a modified version of this objection which points out that the analogical inference to other minds seeks to show, in effect, that I am not unique. But, it is claimed, one needs to know this to be entitled to rule out the fact that differences between me and others are irrelevant, so the analogical inference to other minds begs the question. Again, one may characterise any analogical inference as seeking to show that the sampled cases are not unique. So the modified version fails also.[10]

It is time to look at the objection that the analogical inference to other minds is based on only one case. Although we often demand that an analogical inference be based on more than one instance, nevertheless there are analogical inferences which we take to establish their conclusions with very high probability which proceed from only one examined case. I shall argue that the analogical inference to other minds is one such inference. My argument will proceed via a discussion of an analogical argument which, as it stands, requires to be based on more than one instance.

Suppose I drop an egg on the floor and it breaks. Am I entitled to infer that the next egg dropped will break? Obviously, we should all be happier with such an inference if it were based on a number of cases of eggs that had been dropped and had broken. Why should we be happier? Because we should have a statistically more significant sample? Surely not, because the total number of eggs there are is so large that a few more cases will not yield any significant increase in the percentage of eggs tested. Surely the answer is that if we carry out more tests we rule out the possibility of the breaking in the case of those eggs that have been tested (dropped) being due to something other than their being dropped. What appears to follow from this is that if we know, somehow or other, that the first egg broke just because it was dropped, then the additional tests would not have been necessary. We shall put this point more generally. Suppose that we are arguing from the fact that x1,...xn are both F and G, and that y is F, to the conclusion that y is G. When do we need n to be large? The suggestion we are making is that we need n to be large just when (i) a large sample will provide us with information about the (causal) relation between an x being F and G, (ii) this information is necessary for the inference to y being G, and (iii) we do not already have this information. The important point is (iii). If we already have the required information the sample need not be large.

I can now apply this suggestion to show that the analogical inference to other minds does not need to proceed from more than one case. In Chapter Three I outlined the steps involved in reaching other minds. We do not have direct knowledge that they have (or do not have) a causal role in our behaviour. That is a matter of inference, based on the evidence that is available to us. Given that evidence, we are faced with a decision about the causal role of our mental states. Such a decision, however, is irrelevant to the other minds problem. It is not what enables us to reach other minds. A decision that mental states have a causal role would not help. The relevant principle would be dubious, arguing as it would be from what causes our behaviour to what causes the behaviour of others, and so depending on the principle that like effects have like causes. Happily, that is not how we reach other minds. We reach other minds by inferring in our own case that our mental states are produced by (certain of) our physical (brain) states and then inferring that it will be the same with others. Similar physical (brain) states will have the same consequences. The relevant principle is benign; not that like effects have like causes but that like causes have like effects. In effect, this is an appeal to the Uniformity of Nature.

The crucial move in these steps to reach other minds is that *our* mental states are caused by (certain of) our physical states. That is the relevant causal link. That is known in our own case. Enlarging the sample would not strengthen the inference to that link holding between physical states and mental states. It is, after all, not a matter of observation that mental states are caused by physical states. It is a matter of inference. What is at the centre of the analogical inference to other minds is the causal link between certain physical states and mental states. That is what we need. That can be obtained in my own case. That case provides a multitude of relevant correlations.

The approach to other minds (and the defence against the one case objection) outlined was initially set out in my 'Other Minds as Theoretical Entities'.[11] This represented a significant departure from the approach outlined before that in A. Hyslop and F.C. Jackson 'The Analogical Inference to Other Minds'.[12] The stress has switched from the causal role of mental states to the causal provenance of mental states. Christopher S. Hill[13] has strongly defended the analogical inference to other minds against the charge that its inductive base is too weak. That defence overlaps with both the approach outlined above and the (superseded) approach in Hyslop and Jackson. Hill stresses the importance of biological kinds and I see that as fitting in with my appeal to the uniformity of nature. However, Hill's formulation of the analogical argument and his subsequent discussion seem to be subject to the criticism that the inference being defended is that from similar effects to similar causes. Hyslop and Jackson were certainly guilty of this charge. Hill seems to be stressing the causal role of mental states.

VARIOUS OBJECTIONS TURNING ON UNCHECKABILITY

In this section I criticise objections to the analogical inference to other minds which turn on the fact that it has a logically uncheckable conclusion. But before doing this I argue that what might be called the standard reply misses the point at issue.

The standard reply to the objection that the analogical inference to other minds is an analogical argument with a logically uncheckable conclusion is that there are perfectly acceptable analogical arguments having just this characteristic; namely, analogical arguments from the present to the past.

The crucial error in this reply is that an advocate of the analogical inference to other minds cannot allow that the past is uncheckable in the sense in which they hold that other minds are uncheckable, because the sample class from which the analogical inference constructs the inductive generalisation to other minds *itself contains past events*; namely, *remembered* correlations between behaviour and mental states. Therefore, when it is said, as it so often is, that the analogical inference to other minds proceeds from what we are (directly) aware of, this must be understood as "what we are, or remember having been, aware of". But it is true of any past event that it is logically possible that we should remember being aware of it, thus a defender of the analogical inference to other minds must allow that we might be aware of the past in the relevant sense of "aware".

The various objections that I shall consider in this section have in common the claim that

(P) No analogical (inductive) argument which proceeds from observed facts to a logical uncheckable conclusion is a good analogical argument.

The differences between them lie in the reasons, if any, offered in support of the principle (P) embodied in this claim.

Two points of clarification need to be made about (P). The first is that, as the mental states of others are, strictly speaking, not unobservable (they are whether or not otherwise unobservable, clearly observable by the others whose mental states they are), (P) should be read as prohibiting a particular person's arguing analogically from what they as propounder can observe to what they as propounder cannot observe. I shall follow the usual practice of keeping this implicit, rather than making it explicit, in my discussion.

The second point of clarification concerns the relevant sense in which the analogical inference to other minds is an argument to what cannot be checked. It is commonly asserted that what generates the problem of other minds is the impossibility of being *directly aware* of the mental states of others. As we saw in chapter one, this is a mistake. First, it is not at all clear that it is impossible to be directly aware of another's mental state. Secondly, and more importantly, being directly aware of someone else's pain would not, in itself, tell me that that person was in pain, because I should still require grounds for supposing that what I was aware of *was* someone else's pain (as well as my own). These grounds would have to lie in that person's behaviour (or brain states, etc.). Therefore, the relevant sense in which the analogical inference to other minds is an argument to what cannot be checked is that it is an argument to what cannot be *directly known*. However, nothing in what follows turns on the precise sense in which the mental states of others cannot be directly checked, verified, observed, or whatever.

First Objection

The first objection that I consider consists, in effect, of the bald assertion that the analogical inference to other minds violates (P). Thus, it is urged that the analogical inference to other minds is a bad argument because it purports to establish, on analogical grounds, conclusions about items, namely the mental states of others, about which we cannot have direct knowledge.

Clearly this first, bald objection is a good objection only if (P) can be asserted without offering specific reasons for (P); that is, it is a good objection only if (P) is, in some sense, self-evident. I shall argue, (i) that not only is (P) not self-evident, but that, in the absence of positive reasons for accepting it, (P) should be taken to be false, and (ii) that a possible reason for supposing (P) true involves confusing the possible with the actual.

(i) It is clear for two reasons that (P) is not self-evident. First, the fact that an inductive argument from premises X to conclusion Y violates (P) shows, in itself, only that Y is such that certain kinds of additional evidence for it cannot be obtained. This is to say something which, on the face of it, is quite different from saying that X does not support Y. Therefore, no claim that the former implies the latter can be regarded as self-evident.

Secondly, there are perfectly acceptable inductive arguments which, while not violating (P), violate a principle sufficiently close to (P) for the onus of proof to lie with a defender of (P). The Helium argument discussed earlier in this chapter is such an argument.

That argument proceeds from the fact that certain samples of helium, that have been observed, on earth, have a certain spectrum, to the conclusion that the helium on the sun, which has not been observed, also has the spectrum. Moreover, it is an argument whose conclusion is uncheckable, for it is *impossible* to observe the helium on the sun. Now, of course, this impossibility is empirical, not logical, so there is a difference between the argument for helium on the sun and the analogical inference to other minds. At the same time, it seems clear that this difference is not *obviously* relevant, therefore the onus of proof lies with an advocate of (P); that is, (P) is not self-evident. Indeed, the helium argument surely shows not just that (P) is not self-evident, but that (P) should be taken as false in the absence of positive reasons for (P).

(ii) The argument for the claim that a possible reason for supposing (P) true involves a confusion between the possible and the actual follows Carl Wellman, "Our Criteria for Third Person Psychological Sentences", where he says:

> ... it is easy to see why one is reluctant to accept indirect (analogical) justification where direct justification is absent. There is always a risk in indirect justification, for it is essentially the application of a generalisation based upon past experience to a new situation ... Although the possibility of error cannot be denied, it is reassuring to know that there are tests that could determine whether or not an error had been made. And if there were no possibility of direct verification, then one could never be quite sure.
>
> All this ... is true ... (it) is also irrelevant. Until one has actually obtained this direct verification, it can hardly be included in one's reasons for holding the statement to be true. ... Whether a person might have had or could obtain another (direct) kind of justification is irrelevant to the validity of the justification he actually possesses.[14]

Wellman takes the argument just quoted to refute this first objection to the analogical inference to other minds. It does not. What it refutes is any argument in support of this objection along the lines that any indirect analogical argument is problematic, is taking a punt, in that it involves, obviously, a generalisation from what has been directly verified to what has not been directly verified. However, this problem is resolved by appealing to the *possibility* of directly verifying or failing to verify what one is indirectly arguing to, a possibility which, of course, does not exist in the case of the analogical inference to other minds.

In addition, it is worth noting that those who find reassurance in the possibility of directly *verifying* the conclusion of some analogical argument appear to have overlooked the fact that the possibility of directly verifying the conclusion of such an argument goes hand in hand with the possibility of directly *refuting* the conclusion. (Though it is not true *in general* that direct refutation goes hand in hand with direct verification: witness "I am immortal".) Because the analogical inference to other minds violates (P), it is an analogical argument such that direct observational evidence in favour of its conclusion is impossible. But, it is equally the case that because the analogical inference to other minds violates (P) it is impossible to have direct observational evidence *against* its conclusion. Therefore, if the impossibility of direct evidence in favor of its conclusion counts against the soundness of the analogical inference to other minds, then, by parity of reasoning, the impossibility of direct evidence against its conclusion counts in favor of that argument's soundness. Therefore, the two balance out and leave the soundness untouched.

I now turn to objections which differ from the first in that specific reasons are offered in support of the claim that, because the analogical inference to other minds violates (P), it is unsound.

Second Objection

This is put by Don Locke in this way

> ... if, in an ordinary case, we are doubtful about whether an analogy holds we can check up and see whether it does, but the whole point about other minds is that we can never test whether the analogy does hold.[15]

In effect, Locke's suggestion is that the reason why the violation of (P) matters is that, if we do have doubts, it is impossible to resolve them.

Locke does not make it clear whether the doubt he is discussing is about the truth of the conclusion of the analogical inference to other minds or about its inductive soundness. If we understand being doubtful about whether the analogy holds as being doubtful about whether the cases are sufficiently similar for an analogical argument to be justified then the objection rests on a false premise. The analogy in question is that between one's own behaviour and the behaviour of others, and it is just false that one cannot resolve doubts by checking up on the extent to which one's own behaviour is similar to that of others. On the other hand, if we understand being doubtful about whether the analogy holds as being doubtful about whether others are similar to ourselves in having minds, then the objection is either irrelevant or a *petitio principii*. It is irrelevant if the doubt in question is *psychological*. The fact that if one were to have a certain kind of doubt this doubt could not be removed has nothing to do with whether one *ought* to have this doubt. On the other hand, if the doubt in question is *rational*, the objection is a *petitio principii*. It is the contention of defenders of the analogical inference to other minds that it is *not* rational to doubt whether others have minds just because that argument is a good argument. Therefore, it is blatantly question begging to assume that it is rational to doubt whether others have minds in advancing this objection.

Locke responds[16] to this response to his objection by, in effect, turning that response on its head. He argues that

> the fundamental flaw in the analogical inference to other minds as a response to other minds scepticism, is that it begs the question (p. 155).

Locke gives the following as an instance of the analogical inference to other minds:

> When someone stands on his toe, his response is very much like mine when someone stands on my toe;
>
> Therefore, when someone stands on his toe it hurts him very much as it hurts me when someone stands on my toe.

This (deliberately) schematic instance of the argument, though different from the version of the argument I am defending, is acceptable for present purposes.

He then argues, concerning this instance, that

> the conclusion follows only on the assumption that, to put it metaphorically, where cases are alike externally, they will be alike internally as well. And this is to assume just what the other minds sceptic is questioning. (p. 155)

Locke can, however, only mean by 'follows', in 'the conclusion follows only on the assumption ...', that the conclusion is *well supported* only on the assumption, not that the conclusion deductively follows only on the assumption; for it is an avowedly non-deductive argument that is in question. Hence, the crucial assumption is not (following Locke's metaphorical way of putting it) that *all* cases which are alike externally *are* alike internally as well - an assumption which would in the context of the analogical inference to other minds beg the question. Rather, it is that external similarity provides, in the appropriate circumstances, *good reason* for supposing internal similarity. And this latter assumption does not beg the question against the other minds sceptic. He can happily grant it and still reject the analogical inference to other minds by urging that the circumstances are not appropriate: on the ground, for example, that only one case, one's own, is directly observed. The assumption in question is one which anyone who accepts the traditional inductive practices must accept, hence it is not question-begging against *other minds* scepticism. Of course, the sceptic about *induction* will not grant it, but the analogical inference to other minds has never been intended to be an argument against this more radical scepticism.

Locke also argues, in support of his charge of question-begging that

> what is at question is whether there is a certain analogy between myself and others. It hardly seems appropriate to resolve this question by appeal to an argument from analogy. (p. 154)

There is a fatal equivocation here. Again following Locke's metaphorical mode of expression, 'analogy' either means 'alike externally' or 'alike internally (as well as externally)'. On the first meaning, it is *not* true that 'what is at question is whether there is a certain analogy between myself and others' - *that* is what is granted. And the question is whether this is sufficient, with other available premises, to support the claim that others have minds. On the second meaning, where to assert an analogy between myself and others is to assert an internal likeness as well as an external one, whether there is a certain analogy is what is at issue, but the analogical inference to other minds does not appeal to this analogy. It appeals rather to the external similarity; the internal similarity is what it argues *to*, not what it argues *from*.[17]

Third Objection

This turns on the claim that any acceptable analogical argument must rest on an *independently* established correlation, and that, because the analogical inference to other minds violates (P), there is no appropriate independently established correlation for it to rest on.[18]

Despite the widespread acceptance of this kind of objection, it is hard to see any force in it. It is just false that

> there could be no way of establishing the necessary correlation between the overt moves and their hidden causal counterparts,[19]

because one can establish a correlation in one's own case. It might be replied that a correlation in one's own case between behaviour and mental states could only entitle one to argue about one's own mind, and that to argue to the minds of others it is necessary to establish a correlation between the behaviour of others and the mental states of others. This reply is, however, quite unacceptable; because it amounts to the claim that if x is inferred from y then x must on at least one occasion have been directly observed to be present when y was present. We could not, that is, infer to a party in a shuttered house unless we had on at least one occasion observed directly a party in *that* house, even though we had observed a score of parties in a score of houses other than the house in question.

At this point a modification might be suggested, *viz.*, to construe this third objection so that it turns on the fact that while in normal cases conclusions arrived at analogically can be independently checked, in the case of the analogical inference to other minds no conclusion reached by the use of this method of inference can ever be independently (directly) checked. In short, the modified objection is that the analogical inference to other minds is a method of inference whose use has never, and can never, be shown to be successful.

My reply to the modification follows Hampshire in 'The Analogy of Feeling' where he says

> ... each one of us is constantly able to compare the results of this type of inference with what he knows to be true directly and non-inferentially; each one of us is in a position to make this testing comparison, whenever he is the designated subject of a statement about feelings and sensations (made by another on analogical grounds.)[20]

Malcolm, of course, and his view is widely accepted, claims that Hampshire is guilty of an *ignoratio elenchi* (presumably he meant to say *petitio principii*), arguing that

> ... the reasoning that he (Hampshire) describes involves the assumption that other human figures *do have thoughts and sensations; for they are assumed to make inferences* about me from *observations* of my behaviour[21]

and, obviously, a defender of the analogical inference to other minds is not entitled to make this assumption. Now it is not clear whether Hampshire's argument, as he formulates it, makes this assumption; but it is clear that it need not make this assumption, because it does not depend on the assumption that other human figures are minded. The argument is hypothetical in character; it is that *if* other human figures make inferences about my mind on the basis of observing my behaviour, then those inferences, involving analogical reasoning, can be directly tested.

A more extreme modification of the third objection is sometimes put in terms of the impossibility of knowing that the method of inference involved is reliable; that is, is a method which regularly yields true conclusions. Thus V. Chappell's imaginary sceptic demands that the method of inference be known to be perfectly reliable, where

> ... to know that a method of inference from x to y is perfectly reliable is to know that x occurs whenever y occurs.[22]

This version fails for two reasons. First, the objection depends on an illegitimate identification of a sound inductive method of inference with a perfectly reliable method of inference. The whole point about good inductive arguments, as opposed to valid deductive arguments, is that their premises may be true when their conclusions are false. Secondly, and more importantly, since any defender of the analogical inference to other minds does, of course, believe that their method of inference is very probably, though not necessarily, perfectly reliable: this version of the third objection is a *petitio principii.* Just because a defender of the analogical inference to other minds thinks it is a good inductive argument, they hold that we do know that the method of inference in question is perfectly reliable. The objector can only hold that we do not know with high probability that the method of inference is perfectly reliable, if they assume that the analogical inference to other minds is not a good argument, which is to beg the question at issue.

A possible response here is to insist that a method of inference has to be *shown* to be reliable. We have to see that it works. Now this is possible in one case, *viz.* the hypothetical reasoning outlined above and tested on me. To retort that one instance is not enough is to move back from the uncheckability objection to the generalisation from one case objection that has already been discussed. However, the fatal flaw in this response is that it replaces an old (tested) argument to other minds with a new argument. The new argument in effect says that the old argument to other minds has been successful on a number of occasions, therefore, it can now be used (relied on). But the objection now devours itself. Is this (new) method of inference reliable? Has it been shown to work? If it has, indeed, been shown to work then that, in turn, produces a different argument again, a new method of inference. Has this been shown to be reliable? And so on, and so on. We would never be entitled to use any form of inference were we to accept this principle.

Fourth Objection

This is that because the analogical inference to other minds violates (P), the *observational* evidence on other minds is simply that others do not have minds. As Alexander puts it:

> In spite of the fact that similar behaviour to that I exhibit, accompanied by my experiences, is exhibited by everyone else unaccompanied, as far as I can observe, by experiences - in spite of the fact that is, that I observe more samples of purposive behaviour *without* experiences than I do samples of such behaviour with experiences - I am yet content to conclude that all those samples are really accompanied by experiences, when I should, on the evidence, it seems, conclude that I am unusual.[23]

Now it is, of course, true that when we fail to observe that something is the case this can generally be taken as counter evidence to the claim that it is the case. Nevertheless, as Plantinga points out:

> There is a difference, in general, between failing to observe the presence of A's and observing the absence of A's (observing that no A's are present).[24]

And where it is impossible to observe the absence of A's then the consequent failure to observe the presence of A's is in no way counter-evidence to the presence of A's. Therefore, since it is not possible to observe the absence of mental states in the case of another human figure, our failure to observe their presence is not counter-evidence to there being such mental states.

John Wisdom makes certain comments in 'Other Minds III'[25] on the analogical inference to other minds. These comments could be thought to constitute an "objection" to that inference, though that objection, as would be expected, is far from categorical. It seems to be a variant of this fourth objection. I argue that the variant fails also.

Wisdom writes of the claim that we can establish the existence of invisible germs by analogy with cases of visible germs:

> But the idea which is probable is the idea that there is a *visible* thing, *that* is what analogy suggests. When we have looked and found there isn't any such thing, then that idea can no longer be said to be probable, but is on the contrary known to be false, the analogy is known to have broken down. (p. 69)

But that is too harsh. What we are asked to see is:

> how unlike this ground by analogy is to the ground by analogy for expecting a visible germ which we haven't yet seen, and how like this *association of x's visible, or, if not, invisible with y's* is to the *breakdown of the association of x's with y's*. And grasping the idiosyncrasy of the logic of the argument to invisible x's we shall see how it is that to reply to 'How do I know there are invisible x's?' with the response 'By analogy' is to reply 'In the way you do'. (p. 71)

Essentially, such analogical inferences are:

> uncommonly like those cases where we say, 'But it was found that there the analogy breaks down'. (p. 71)

Finally, his comment on the analogical inference to an invisible measles germ is that:

> before we have looked and found no germs *then* the position is like that which we describe as knowing by analogy that this animal will have big teeth (by analogy from a smaller animal with small teeth). But after we have looked, *then* the position is very different, and to

> describe it also as knowing by analogy that there are germs seems to near to saying that when we find that the enormous animal has no teeth, what we find is that it has invisible teeth. (p. 74)

Now it is not clear what Wisdom draws from these points nor even what he wants to draw. In particular, it is not clear *what* he thinks the justification is for our belief in other minds though it does seem that he thinks he has described that justification without seeking to characterise it. However, his view is obscure, apart from our being informed "that there are more logics than are dreamt of in the textbooks" and that we know "in the way we do" (p 71). What seems available to him would be the suggestion that the other minds belief involves an inference to the best explanation grounded on the analogy between each of us and other human figures. In his view, to persist with the straightforward analogical inference would lump one with the Believer in Wisdom's 'case of the invisible gardener' where a belief is defended by progressive refinement against whatever unfavourable evidence is produced.

Since I am not concerned here with Wisdom's own solution, if any, to the other minds problem, it is this latter view that will be examined.

Take the three analogical arguments discussed by Wisdom: from small teeth to invisible teeth, from visible germ to invisible germs, and the other minds inference. The first, to invisible teeth, parallels the invisible gardener. The second, on the information supplied, is best treated as an inference to the best explanation. However, it could be defended as a straightforward analogy if there had been observed a gradation of germs more and more difficult to detect and analogy suggested that a malarial germ, if it existed, would be empirically unobservable. (The case where an inference was drawn to a germ logically impossible to observe would be quite different for it is completely obscure how the analogical warrant could be provided for such an inference from visible germs.)

The third argument, the analogical inference to other minds, is a straightforward analogy and is defensible. Wisdom explicitly affirms, in the teeth case, that large visible teeth are probable *before* we have looked, and in the germ case, that what is probable is a visible measles germ. He seems, that is, to affirm some such principle as: where y is inferred by analogy from x and x is visible then y is visible too. But this principle seems in need of supporting argument for x and y obviously do not need to have all their properties in common; and, indeed, if they had all their properties in common, we should be involved in the identity of indiscernibles problem. The most that one can justify as a general requirement is that there be *some* properties in common. There seems no *prima facie* reason for singling out observability as a property which, if it belongs to x, necessarily belongs to y. If one's inference were to a visible germ then failure to observe that germ would be counter-evidence. My claim is that it simply does not follow from the fact that y is inferred from x and that x is observable, that the inference is to an observable y.

There is, even so, a sharper counter to Wisdom. By analogy with the case of the missing dentures and the case of the elusive germ, what is probable given Wisdom's 'principle', *before* we look for other minds, is mental states visible to their owner alone. As he writes elsewhere:

> The peculiarity of the soul is not that it is visible to none but that it is visible only to one.[26]

and just *that* is what he ought to say is rendered probable by analogy. If this is so, failure to observe those mental states is not counter-evidence. But in fact, though we infer by analogy to the existence of mental states other than our own we do *not* infer by analogy that those mental states will be observable only by their owner. The claim that they are so observable is justified by the claim that it is a necessary truth about mental states that they are so observable. Unless it is claimed, most implausibly, that the user of the analogical inference has no right to this knowledge, what is probable in the other minds inference *before* we look is that the mental states in question will not be observable by the inferrer. Given this, the failure to observe the mental states is not counter-evidence to the inference. Indeed given what is possible before we look the sensible analogical arguer will not bother to look.

Fifth Objection

This objection is put forward by A. Plantinga.[27] It does not turn on (P) but does however turn on a principle akin to (P), namely (A), below.

Plantinga's objection may be set out as follows. "A simple inductive argument for S" is defined as an argument of the following form:

> Every A such that S has determined by observation whether or not A is B, is such that S has determined by observation that A is B. Therefore probably every A is B.

Plantinga then suggests that

(A) A simple inductive argument for S is acceptable only if it is logically possible that its sample class contain a counter instance to its conclusion.

Plantinga's objection is that the analogical inference to other minds requires certain simple inductive arguments that violate (A). For example, it requires the inference from

(1) Every case of pain behaviour such that I have determined by observation whether or not it was accompanied by pain in the body displaying it, was accompanied by pain in that body

to

(2) Probably every case of pain behaviour is accompanied by pain in the body displaying it.

And the inference from (1) to (2) violates (A) because it is logically impossible to get in the sample class a counter-instance to the conclusion. This is logically impossible because I cannot determine by observation that a body does *not* contain a pain; I can only determine that it does not contain a pain of *mine*, which leaves open the possibility that it contains a pain belonging to someone else, as any body, including my own, may contain someone else's pain.

I agree with Plantinga in accepting (A), and agree that the inference from (1) to (2) violates (A). However, the analogical inference to other minds doe *not* involve an inference from (1) to (2). Now, it is easy enough to show that it does not involve an inference from (1) to (2), by, for example, pointing out that any analogical arguer must regard (2) as false in virtue of the phenomenon of pretending. However, it is clear that this does not get to the heart of the matter. What is needed to show Plantinga's objection fails is an inference which, while being of the same general character as that from (1) to (2), does not violate (A). I suggest that the inference from

(3) Every case of pain behaviour such that I have determined by observation whether or not it was accompanied by a pain in

the body displaying it *which belonged to the person whose body displayed the pain behavior,* was so accompanied.

to

(4) Probably every case of pain behavior is accompanied by a pain which is in the body displaying it and which belongs to the person whose body is displaying it,

is such an inference. It is an inference which constitutes a simple inductive argument, and it does not violate (A), because it is possible to determine by observation that a body does not contain a pain of mine.

A possible reply[28] for Plantinga at this stage, is to point out that the step from (3) to (4) does not violate (A), only if we are allowed to talk of the person whose body displays the pain behaviour in (3). If we allowed (as Plantinga no doubt holds to be possible) that human bodies may have more than one person "attached" to them, and, so, talked merely of a person in (3), then a counter-instance to the conclusion would not be possible. A counter-instance would not be possible, and, so, (A) would be violated, because if I determined by observation that my body was displaying pain behavior and that my body did not contain a pain of mine, it would still be possible that my body contained a pain of a person (other than myself) who shared my body with me.

My counter to this reply is that, though I agree both that the assumption that there is no more than one person attached to my body is not necessarily true, and that the assumption is needed to allow us to replace (1) by (3), and so to avoid violating (A), nevertheless the assumption is *true*. And if the assumption is true, and known to be true, we are entitled to make it in (3). It would of course be circular then to use the analogical argument to support the assumption, but it is obvious in the first place that one could not expect the analogical argument to establish that there is just one person to my body because the analogical argument proceeds from what I am directly aware of, and I am not directly aware of this fact.[29]

Karl Ameriks[30] has added, in discussing Plantinga, another principle, which he claims has to be met if an analogical argument is to be rationally acceptable, and, he would claim, (3) fails to meet it. Ameriks takes an analogical argument to have this general form: (practically) all

A's which are B's have C's, hence probably any non-A which is B has C. The additional principle is this:

> Something's being a C does not make it *prima facie* likely that it's an A rather than a non-A.

Ameriks argues thus for this principle:

> It is necessary because otherwise it would be unlikely for something which is not an A to be a C, and so it would not be probable that non-A's which happen to be B's are C's. Thus we would be foolish to infer from the fact that all unmarried males are bachelors to the fact that most unmarried females are bachelors.[31]

We can ignore the fact that only unmarried males of marriageable age are bachelors, and point out that this example, and the others he uses, involve entailment. It is entailed, not *prima facie* likely, that a bachelor is an unmarried male rather than an unmarried female. Elsewhere (p. 288) he talks instead of what is *a priori*, which is, of course, something else again. The relationship (as discussed by Ameriks) relevant to other minds does not help, *viz.* between (i) being sensed as the location of my pain and (ii) being a state of my body. (i) does not entail being (ii). Nor is it an *a priori* truth that (i) involves (ii). Nor is it the case that what is (i) is *prima facie* likely to be (ii). The relationship between (i) and (ii) is contingent (certainly no analogical arguer should think otherwise). There is more to what makes a body mine than the location of (some of) my feelings. That aside, it is contingent which body (or bodies) is mine. (3) survives Amerik's strictures.[32]

Sixth Objection

The last objection I consider is that put forward by Sydney Shoemaker. He says:

> It has often been suggested, of course, that one can discover empirical correlations between *one's own* psychological states and facts about one's own body and behaviour, and that one's knowledge of these correlations can provide a basis for inductive inferences from the bodily and behavioural states of others to assertions about their psychological states. ... But even if one could discover this, one

> could have no grounds for supposing that the correlations that hold in one's own case hold in the case of other persons as well.[33]

Shoemaker's reason for holding that

> one could have no grounds for supposing that the correlations that hold in one's own case hold in the case of other persons as well

is that anyone who employs the analogical inference to other minds presumably holds that behavioral and psychological states are only contingently related; and, therefore, that it is contingent that the relation which holds in one's own case holds in the case of others. But, if this is contingent, the analogical arguer can have no ground for supposing it to be the case, because the only empirical evidence he has concerns the relation in his own case. He has not, for example, independently established a correlation between the relationship between physical and psychological in his own case and the relationship in another's case, to sustain the generalisation that the relationship is the same in all cases.

Shoemaker's objection raises issues in inductive logic which are beyond the scope of this chapter. I shall content myself with pointing out that if his objection is successful, then *no analogical arguments whatever are successful;* because what Shoemaker says about the analogical inference to other minds can be said about any analogical argument concerning the unobserved. Whenever one argues inductively from all observed As having been Bs to the unobserved As being Bs, it will be true both that it is contingent that the unobserved As are like the observed As (otherwise the argument would be deductive, not inductive), and true that there is no independently established correlation between observed As and unobserved As to sustain the generalisation that observed As and unobserved As are alike (otherwise the unobserved As would be observed, not unobserved).

It seems to me the analogical inference to other minds survives the best that these various objections to its inductive soundness can manage. I commend it to you. It can be used with complete respectability. I am sure it is used.

CHAPTER FIVE

CRITERIA AND OTHER MINDS*

The analogical inference to other minds has competitors other than the scientific inference to other minds. They all seem to me to fail. I shall examine them successively beginning with that "approach" to other minds which depends on the notion of criteria. The classical role of criteria in epistemology has, indeed, been to justify belief in other minds. Often this has proceeded in two stages. First the analogical inference to other minds is attacked and thoroughly discredited thereby exposing a vacuum. Criteria are then invoked to fill this vacuum. Given that there is such a strategy, the consequent urgency of the need has, I think, been responsible for the fact that criteria have been rushed in without due care and attention. This chapter will give them that due care and attention.

Criteria have also been used in the context of other minds to provide an account of what is meant by third-person psychological statements, an account which proceeds in terms of other than necessary and sufficient conditions. There is notoriously a problem in reconciling the meaning of such third-person statements with the meaning of first-person psychological statements, given that this account has behaviour (in certain circumstances) as central. The problem continues to seem insuperable.

However, the epistemological role of criteria, its use in justifying belief in other minds, is also often associated with moves to reject the problem of other minds by way of rejecting the claimed epistemological asymmetry between our own case and that of others, sometimes through the claim that we have direct knowledge in our own case. I shall have something to say about that "symmetry" issue in the next chapter. In the meantime, I shall just make the point that, against such a "backdrop", it is not surprising that the criterial "solution" that is often, perhaps unwisely, tacked on, is generally unpersuasive since there will hardly be a solution in the absence of the problem. What such criterialists are doing is perhaps best viewed as epistemological anthropology, describing how human beings, as a matter of psychological fact, treat other minds. Once again, there will be more of that in the next chapter.

However, my interest in this chapter is in whether criteria can, indeed, solve the problem of other minds. They have, indeed, been used in order to do so. In such cases that there is a problem of other minds is conceded. However, such criterialists may be seen as rejecting in some way my characterisation of the problem. They see our knowledge of other minds as, in some way or other, though not direct, nevertheless non-inferential. It is not based on inductive inference.

Since the justification which criteria afford is claimed to be non-inductive, it is, therefore, different from that afforded by the argument from analogy for the existence of other minds (and from the scientific inference argument for other minds). Generally, it is claimed that criterial justification is *strong* justification, where this is thought to distinguish it from the analogical inference. However, that criterial justification is indeed strong justification is almost invariably just assumed. Where it is claimed to be part of the definition of 'criterion' that criterial justification is strong justification, it is simply assumed that criterial justification is available in the other minds case.

If criteria are to be used to support belief in other minds, then the following conditions must be met by any such approach to the problem of other minds: (i) in general, the behaviour (in certain circumstances) of other human figures should be a criterion for their being minded; (ii) the justification afforded should not include, as an essential element, an appeal to an analogical inference or, indeed, any inductive inference; and (iii) our actual beliefs about other human figures must be supported to the required extent. For our belief about each human figure that we come across, or read about, or whatever, is that the human figure has thoughts, emotions, experiences etc. and we hold this belief as strongly as any belief can be held. This chapter argues that the accounts of criterial justification for other minds proffered in the literature do not meet these conditions; i.e. each of the various cases made out under some specific sense of 'criterion' fails to meet at least one of these conditions. The three conditions seem non-contentious and are implicitly accepted by all those who rely on the use of criteria to justify belief in other minds. My central concern in this chapter will be to examine the relevance to the other minds problem of that view of a criterion which sees it as evidence which is necessarily evidence.

Two general, preliminary points can be made. There is a debate in this general area which revolves around the question whether an inner process needs an outer criterion (a more particular focus for this debate is the issue of a "private" language, to be discussed in the next chapter). But the problem in the case of other minds is to move from what we have direct knowledge of, to that of which we lack direct knowledge. That an inner (mental) process needs an outer (behavioural) criterion is not, of itself, directly relevant to this problem, for the movement we want is in the opposite direction. The fact, if it be a fact, that if we itch we scratch, or want to, is *prima facie* irrelevant to that movement. That we need an outer criterion if talk of inner processes is to be meaningful, does not give us what we want. At best, it would provide an essential condition for what we want. That the belief is meaningful does not make it justified. What is in question is our entitlement to move from cases of scratching to cases of itching. We want to know how to justify going from scratching to itching, not the converse.

The second general, preliminary point is this: that an inner process needs an outer criterion does not tell us in a particular case what the criterion will be. In the case of other minds, bland assertions that nothing other than behaviour in certain circumstances is plausible as a candidate for the criterion are useless without supporting argument. Behaviour is clearly only one among many observable features of human figures, and observability is the only constraint that the need of an outer criterion entails. That some specific observable features are the criteria for the mental states of other human figures needs support. If behaviour is assigned the key role, we need to know how particular mental states are "married" to particular behaviour.

Too often, these two *prima facie* problems are ignored. Of the two criterial positions I shall examine as representative, the first[1] seems guilty of this charge whereas the second[2] is not. I shall argue, though, that neither position meets all three conditions of success for criterial support.

I

CRITERIAL EVIDENCE AS EVIDENCE THAT ESTABLISHES WITH CERTAINTY

This sense of 'criterion' was made famous by Malcolm. A criterion is something that *settles* a question beyond doubt, and not just in the commonsense notion of certainty:

> Even if physiology has established that a specific event in the brain accompanies bodily pain, still it *could* happen (it makes sense to suppose) that a man might be in pain without that brain event occurring. But if the criterion of being in pain is satisfied then he *must* be in pain.[3]

But despite this, as is well-known, Malcolm does not think that the relationship which holds between behaviour (in whatever circumstances) and mental states is one of logical implication. Nor, apart from the behaviourist, does anyone else. If the relationship is not logical implication then it would seem to be evidential. Here the structure of Malcolm's position becomes crucial. His view is that the analogical inference to other minds is either meaningless or superfluous. However, if Malcolm's view is that it is superfluous, his alternative must do what has to be done by way of justifying our belief in other minds. If Malcolm's alternative provides only moderate evidential support, then the analogical inference to other minds will be needed unless Malcolm's alternative is stronger. The question will be begged against that inference unless argument to this effect is supplied. But all we get in the form of argument for the alternative is the claim that we just *do* accept certain behaviour in certain circumstances as guaranteeing our belief.[4] The only thing that could be said for this by those who think the analogical inference to other minds is a very weak argument is, presumably, that no argument is better than a bad one.

However, there seems to be implicit in Malcolm an argument which would make sense of his talk of certainty and of the dilemma he offers to the analogical inference to other minds (that it is either meaningless or superfluous). The argument runs like this: (1) the analogical arguer knows how to use third person psychological sentences; (2) if they do they have sometimes used such sentences to make true assertions (positive ones); and (3) if so, then other minds exist. This argument is hinted at by Malcolm when he claims:

> At a primitive stage of teaching one does not require or expect definitions, but rather that the child should *pick out* reds from blues, dogs from cats, spoons from forks. This involves his looking, pointing, reaching for and going to the right things and not the wrong ones.[5]

And also:

> The satisfaction of the criterion of *y* establishes the existence of *y* beyond question; it repeats the kind of case in which we were taught to say "*y*".[6]

Its principle is explained in *Dreaming*:

> The absurdity comes down to this, that for the judgement to be *true* the person who made it would have to be asleep ... that there could not be a criterion for the correct use of the words 'I am asleep', depends on that: for to know that a person uses those words correctly we should sometimes have to observe him judging that he is asleep *while* he is asleep.[7]

Given that Malcolm agrees it is logically possible that someone know a language without having been taught it, the real question here is how we know that someone is able to use sentences correctly. There are just two possibilities: either someone wants to know this of someone else, or they want to know this of themselves. On the first possibility, if what is in question is the understanding of third person psychological sentences, then Malcolm's argument is clearly question-begging. If Fred wants to know that George is able to use such sentences correctly, he wants to know this of a person, which would involve knowing that there are other minds. As a counter to this, it might be suggested that Fred wants to know this of another human figure, and will therefore check on assertions made by this human figure about other human figures. However, Fred's check, in order to be relevant to the other minds problem, must be of assertions made about human figures other than Fred. But how is Fred to know whether such assertions are true without a solution to the other minds problem? On the second possibility: (a) it, for the same reason, is question-begging (Fred in that case wants to know whether his use of third-person psychological sentences is correct so on Malcolm's principle he needs to know whether his assertions are true so he clearly needs a solution to the other minds problem); and (b) how can Fred check on his own assertions without knowing in advance what he means by them? So Malcolm is left with his claim that we just *do* accept behaviour in certain circumstances as guaranteeing our belief.

It is worth noting that the two strands I have separated connect very naturally. The claim,

> It repeats the kind of case in which we were taught to say "y"

is capable of being pulled in two directions. One is towards cases in which we have learned that we call this an instance of *y*, so that this is an instance of the kind of thing that is meant by "y". The other is towards the view that cases in which we are taught the use of a term must be authentic cases.

Both strands connect with the two preliminary points about the relationship between inner process and outer criterion. If two human beings are to teach one another a language, given their present powers, they each need some way of picking out the relevant aspects of their experience. The child hurts itself and cries: the circumstances and its behaviour serve to pick out what the word 'pain' refers to, though they are not, of course, themselves that to which the word refers. There is nothing in this that provides a short cut to a justifiable belief in another mind. For (i) that human beings need this aid to communicate does not mean that where there is a language there must be human beings; and (ii) our use of circumstances/behaviour is precisely what we need to justify in the other minds problem; for what entitles us to our belief that there is something other than the behaviour/circumstances involved? This second point is one that the alternative account of 'criterion' purports to meet.

II

CRITICAL EVIDENCE AS EVIDENCE WHICH IS NECESSARILY EVIDENCE

I offer a series of quotations:

(a) Criterial evidence for *p* is that which is evidence for *p* in virtue of the meaning of *p* ... If *x* is a criterion of *p* , then it is logically necessary that *x* is evidence for *p* .[8]

(b) The defining characteristics of mental concept words are of two distinct kinds, behavioural and psychical ... the very nature of mental concepts provides us with a logical justification for ascribing to others subjective experiences.[9]

(c) A phenomenon or circumstance (or set of phenomena or circumstances) which constitutes evidence that the concept in question applies in virtue of the very structure of the concept, and which hence constitutes 'non-inductive evidence' that the concept in question applies.[10]

(d) For present purposes we may characterize the criteria for the truth of a judgement as those states of affairs that are ... direct and non-inductive evidence in favour of the truth of the judgment ... We know that it is evidence, not by having observed correlations and discovered empirical generalizations, but by understanding the concept of a y and the meaning of statements about the identity of y's.[11]

The early criterialist period (Malcolm) can be seen as an unsuccessful search for a middle way between a deductive inference to other minds and an inductive inference. The deductive inference was too strong. The inductive inference was too weak. The high tower of necessary truth was barred. The bottomless pit of scepticism loomed, distressingly near.

The quotations above, individually and collectively, offer the much sought for soft landing for the criterial relationship, in a "halfway house", not deductive but not inductive either, and 'logical' because it is a matter of entailment that criterial evidence is evidence. However, despite this initial promise, there is an initial problem. If we match the characterisation against our third condition it is wholly obscure how it measures up. That evidence is necessarily evidence does not entail that it is 'first-rate evidence',[12] nor, indeed, that it occupies any particular point on the scale running from strong through to weak. Coburn[13] says that it may or may not be conclusive evidence, but he does not indicate what it might reduce to if it is not conclusive evidence. I know of no argument to the effect that what is necessarily evidence is thereby good evidence.

There are two ways of applying this "necessary evidence" characterisation of criteria to the other minds problem. Since what is required is an inference from the behaviour of others to the experiences of others, if this inference is to be criterial in the sense just outlined, then one way of achieving this will be to claim that it is a necessary truth that, in general, psychophysical relationships are as they are. It does not look to be a necessary truth, for a natural response to make to the claim that the behaviour of others is necessarily evidence of their experiences (via the necessary truth invoked) is that I can imagine my psychophysical correlations altering. Of course, what I can imagine is here as elsewhere not really the point. Some have nimbler imaginations than others.

Intuitively, psychological relationships smack of contingency. That is the instinctive response.

The claim is discussed and countered in the literature and we find there two arguments to the effect that in one's own case it is a necessary truth that the psychophysical relationships are as they are. Both are found in Shoemaker.[14] The first depends crucially on the view that

> The claim that one can imagine what it would be like for the state of affairs to become, in one's own case, the rule rather than the exception is, I think, equivalent to the claim that one could *establish* that this had occurred in one's own case ...

This view is discussed and (rightly) rejected by Plantinga[15] and by Williams.[16] The second is set out by Shoemaker[17] and is discussed and accepted, though in modified form, by Lycan.[18] Both arguments are to the effect that the knowledge I have of my psychophysical relationships is not empirical. The point I want to make applies against both. For the conclusion of both arguments is that it is a necessary truth that *my* psychophysical relationships are as they are. But from this conclusion nothing follows about the psychophysical relationships holding for other human figures. The result of the criterial move is simply this: where the analogical inference to other minds relies crucially on a contingent first-person premise the criterialist claims to have a non-contingent premise. True, the evidence one has is not then the result of empirical observation. But it remains evidence limited to one's own case and is thus on a par with the analogical inference to other minds in everything relevant to its strength as evidence. The entitlement to infer from the behaviour of others still depends on the correlations holding in one's own case. This result is hardly surprising, since, within the context of this debate, it is non-controversial that it is logically possible that the behaviour of other human figures is unaccompanied by experience.

A possible response here is to claim that every psychophysical relationship is mine for someone. This reply is unacceptable, since it assumes that there are, generally, psychophysical relationships and this is precisely what is at issue in the problem of other minds. In short, the reply would be question-begging.

Another possible response (this time from Shoemaker) is that the traditional analogical inference to other minds is mistaken in its view that its premises are contingent (pp. 191-2). My reply to this is (i) that this, as I have argued, does not affect its worth as an argument; (ii) that if this is all Shoemaker is trying to show then the criterial justification for other minds, as an alternative to the analogical inference to other minds, remains to be set out; and (iii) that it is clear that Lycan's intent is to support the non-contingency *in general* of psychophysical relationships.

The second attempt to treat the behaviour of human figures as being necessarily evidence that they are minded has been set out in a particularly clear form by Aune.[19] He outlines the connection he believes to hold between itching and scratching. Itching does not entail scratching and scratching does not entail itching, but itching does entail being disposed to scratch. Aune claims that, consequently, scratching can legitimately be called 'primitive, unlearned itch-behaviour' and serve as a 'criterion' for the presence of itches. He claims further that a criterion provides 'first-rate evidence':

> The fact, in other words, that certain kinds of behaviour provide first-rate evidence for the presence of a given sensation is not merely a contingent truth but a conceptual necessity: it is, so to say, built right into the concept of the sensation (p. 203).

I think Shoemaker's[20] claim that there are necessary truths of the form

> if someone is in mental state *m* then (normally, *ceteris paribus*, etc.) he will behave in way *B*

amounts to the same position. Shoemaker's explicit reference to Armstrong's causal analysis of the mind, where an itch is (in part) an internal state apt for the production of scratching, would seem to support this ascription.

In this approach to the problem of the existence of other minds, two moves are made. First, the claimed entailment from being in a particular mental state to being disposed to behave in a certain way is used to establish that the behavioural evidence is necessarily evidence 'built right into the concept' and thus not dependent on one's own case. Given that dependence on one's own case is thought to be a fundamental weakness of the analogical inference to other minds, then the role of the entailment, viz. avoidance of that weakness, is crucial. However, and this is the second move, that the scratching is evidence of itching follows from the fact that the itcher is disposed to scratch, and it follows irrespective of whether the fact is necessary or contingent. The principle appealed to here is: where p disposes an x to Φ then Φing is evidence of p (henceforth (R)). The sense of 'evidence' involved is that where to say that p is evidence of q means that q is more probable than not, given p (and nothing else).

Now this principle will not do, for the following reasons.

(i) It is conceivable that Φing be such that both p and q dispose an x to Φ and further that p entails $-q$. By (R), Φing is evidence of $-q$. So Φing is evidence of q and $-q$ which is unacceptable. Consider the case of a car which will neither start without petrol nor start when the carburettor is flooded. (R) would have it that its not starting is evidence for its having no petrol *and* for its being flooded.

(ii) No limit in principle can be placed on the number of states of affairs that would dispose a car not to start. (R) would have it that the car's not starting is evidence for each and every one of those states of affairs. Presumably this would supply us with an excess of reasons for the car's not starting.

(iii) If someone who wonders what is causing an x to Φ knows that p disposes an x to Φ, then they know on this basis only that x's Φing might be caused by p . That it is conceivable that x's Φing is caused by p is no reason at all for saying it is so caused. What we need is information about the circumstances under which Φing occurs, which might then allow us to form an opinion about what is in fact causing the phenomenon in this particular case. Here the analogical inference to other minds can at least point to one's own case as relevant insofar as it can be claimed that in that case Φing *is caused* by p, however unhelpful in the event that turns out to be. And such information is what is needed if this manoeuvre is to have any chance whatever of succeeding; for it is conceivable that scratching is never, in actual fact, caused by itching. Indeed, the following case is conceivable. Itching disposes me to scratch; as soon as I scratch, and, indeed, as soon as I move to scratch, the itching ceases. Here scratching is evidence against the presence of itching during the scratching period.

(iv) This objection is not an objection to (R) as such. It is *ad hominem* against those philosophers who espouse both this criterialist position on the existence of other minds and the claim that you cannot learn from your own case the meanings of mental terms. The criterialist claim is that there is an entailment from itching to being disposed to scratch. Using this entailment, and given (R), the criterialist claims that the behaviour of other human figures is evidence for their being minded. They claim that this is known to be true independently of their own case and thus frees their evidential base from the classical objection to the analogical inference to other minds - that it is inadequate because based on one instance. However, this crucial move depends on the entailment; and knowledge of the entailment depends on knowledge of the meaning of mental terms like 'itching'. Given that this knowledge is crucial to this particular criterialist support for the belief in other minds, it must itself be independent of any appeal to the existence of such minds. This requirement demands that it be known from one's own case, so the criterialist position is inconsistent with the claim that one cannot learn from one's own case the meaning of mental terms (though, as we shall see in the next chapter, it is compatible with the claim that a private language is impossible).

The above objections constitute my case against (R). Before starting on that case, I pointed out that the claimed entailment from mental state to disposition to behavioural state is used by the criterialist to avoid the necessity for an inductive inference based on one's own case. Such a move is crucial, since otherwise the criterial position would be in the same inductive boat as the analogical inference to other minds, and would thus face both classical objections to that inference - that its evidential base is woefully meagre and its conclusion is uncheckable. It might be thought that the criterialist position fails to avoid the objection that its conclusion is uncheckable. However, that objection is to inductive arguments with uncheckable conclusions and the criterialist argument is neither inductive nor deductive. However, some of the "uncheckability" objections discussed in Chapter 4 would seem to apply to the criterialist position.

I have five objections to the claim that the entailment required holds. (i) When Aune claims that the entailment holds he reminds us of Ryle's observation[21] that pain is a sensation one ordinarily minds having:

> ... and any feeling that does not dispose one to seek some kind of relief, to exhibit some kind of avoidance-behaviour, is thus not a pain but something else.[22]

But that there is an entailment to some kind of avoidance-behaviour is not to the point if it is contingent, as it surely is, what kind of behaviour is avoidance-behaviour. Similarly, that there is an entailment to relief-seeking behaviour is not to the point if it is a contingent matter just what behaviour it is that provides relief. What the criterialist must claim is that though it is contingent that scratching relieves itching, it is a necessary truth that if one itches one is disposed to scratch, i.e., it is a necessary truth that one is disposed to that behaviour which as a matter of contingent fact provides relief.

I offer a description of an imaginable state of affairs. What gives relief or pleasure when we have an itch, wherever that itch be, is to place our left hand on top of our head. Whenever we scratch that portion of our body where we itch violent pain ensues. Now it is clear what we in fact would do in such a universe. We would place, not scratch. The criterialist must insist not merely that we, as a matter of fact, *might* be so arranged that we were still disposed to scratch, but that it is *entailed* that even in such a universe we should be so disposed.

Take Hoyle's Black Cloud (referred to as a non-human person by Aune[23]). It is conceivable that the Black Cloud itched. It is implausible to maintain that when it itched it *must* have been disposed to scratch the affected area (where the rubbing the affected area would qualify as scratching in the absence of limbs).

A less fanciful way of making the same point is this. Consider a mild headache. Apart from a disposition to take aspirin - which is not universal, never mind entailed - there are generally no dispositions, apart from the idiosyncratic and contingent, attendant on this irritating but endurable condition. Even a severe headache does not seem to be accompanied by universal dispositions. The point is that relief in the case of a headache is generally dependent on drugs, and there is not anything one can do otherwise to 'avoid' the pain.

(ii) When we itch we say we want to scratch. This is inaccurate. What we really want when we itch is to achieve the immediate effect that rubbing the itching portion of our body achieves. The itch demands, as it were, the sensation of being rubbed. It is contingent that this sensation is achieved by rubbing the itching portion, for it might be the case that one got the sensation of rubbing at x by rubbing at y.

(iii) Philosophers have tended to confine their discussions to pain in general. But there are different kinds of pain, and the differences cannot always be made out in terms of behaviour. Take a burning pain. This is not distinguished by the behaviour to which it gives rise but by the circumstances in which it typically occurs. However, it is not a necessary truth that you get a burning pain when burned; and you can have a burning pain when you have not been burned. A burning pain is, roughly, a pain that is *like* the pain which we get, as a matter of contingent fact, when burned.[24] The exact characterisation of a burning pain is not in question here. What matters is that since behaviour will not serve to distinguish such pains, our immediate criterialist cannot deal with them. It would be blatantly question-begging, even granted a conceptual connection between burning pains and burning, to claim that being burned provides grounds for believing that the burned person is in (burning) pain. The most the conceptual connection could provide would be that *if* there is pain, it will be burning pain.

It may be admitted, as Aune claims, that certain sorts of pain occasion peculiar behaviour:

> ... a sharp pain may dispose a man to jerk or jump, as though he had been jabbed by a pointed object; a beating headache may impel a man to throw his head downward, as though he were continually struck on the head; and a nagging pain may produce those familiar symptoms of extreme frustration.[25]

But three points must be noted. Firstly, it is implausible to claim that the disposition is entailed, and it is significant that Aune himself says 'may dispose'. Secondly, in order for the criterialist to achieve the desired entailment, the kind of pain must be characterisable in terms of its effects, and not all kinds of pain can be. And thirdly, consider the kinds of pain listed by Aune. A nagging pain, whatever its exact characterisation, is not characterised in terms of its effects:

> ... we use words such as "nagging" or "pulsating" to bring out the quality of our pains by analogy with the behaviour of such things as wives and currents.[26]

I think a beating headache is also conceived analogically, while a sharp pain is surely characterised in the same way as 'burning pain': the sort of pain a person (at least in some cases) typically has when they are stabbed or jabbed (see footnote 24).

(iv) Language is thought to be crucial in our knowledge of other minds. That other human figures speak, write, and so communicate with us, seems highly relevant if we are wondering whether they are minded. The criterialist might respond to the previous objection by inserting verbal behaviour in the gap. But the criterialist cannot use the notion of necessary evidence to justify our taking verbal behaviour as evidence, for there is no entailment from thinking that I am in pain to asserting that I am in pain; nor is there an entailment to being disposed to do so. In any case, the crucial move is to treat 'I am in pain' and other utterances, in whatever form, as evidence, and there is obviously no entailment from thinking that I am in pain to being disposed to utter 'I am in pain'.

(v) It is significant that the psychologists distinguish basically between only two kinds of behaviour, learned and unlearned. For them the question is whether scratching, as a response to an itch, is learned or unlearned. For the criterialist the question could not arise. Yet surely it is a subject for empirical enquiry.

So the claimed entailment is to be rejected, (R) has been shown to be untenable, Shoemaker's "necessary evidence" has been disposed of, and Malcolm's use of criteria found not to work. It does not seem that criterialism constitutes the slightest threat to the analogical inference to other minds (pure or hybrid).

It would seem that John McDowell would agree with this claim, though without any commitment to that inference. John McDowell has attacked the notion of criterial evidence in general, and particularly in the case of its use in the context of other minds.[27] The problem he sees in criterial evidence is that criterial evidence is defeasible. Criterial evidence is defeasible in that a belief justified on the grounds of criteria can later be set aside, though the relevant criteria were, indeed, satisfied. Further information can lead to the belief's being set aside. Cases of pretence would be examples of this happening, where the pretence is subsequently discovered.

Care has to be taken in formulating the problem that, supposedly, besets criteria. The intention is to criticise specifically criterial evidence not inductive evidence generally. McDowell's intention is not to criticise, for example, the analogical inference to other minds on the ground that, in a particular application, it is defective because later evidence sets aside the previous belief. I have in mind cases of pretence. More generally, McDowell's target is not all those cases of reasoning where the evidence is compatible with the conclusion's being false. Whether or not that is his intention, such a general aim would, of course, be suspiciously ambitious. What he wants to do, and needs to do, is to drive a wedge between criterial evidence and other inductive evidence. Though that is not sufficient for his attack on criterial evidence, it is clearly necessary.

He rejects the notion that criterial evidence is at fault because it sanctions the idea that one can know that something is the case when it is not the case. As he points out, that is clearly incoherent. That is not what he sees as being what is wrong with criterial evidence. What he does insist on is that criterial evidence is evidence compatible with the falsity of what is claimed to be known, as the following quotations make clear:

> ... even on the occasions that seem most favourable for a claim to be able to see that someone else is in some 'inner' state, the reach of one's experience falls short of that circumstance itself - not just in the sense that the person's being in the 'inner' state is not itself embraced within the scope of one's consciousness, but in the sense that what is available to one's experience is something compatible with the person's not being in the 'inner' state at all. (p. 457)

> Nevertheless, the 'criterial' view does envisage ascribing knowledge on the strength of something compatible with the falsity of what is supposedly known. (p. 459)

> How can an appeal to 'convention' somehow drive a wedge between accepting that everything that one has is compatible with things not being so, on the one hand, and admitting that one does not know that things are so, on the other? (p. 458)

It is initially, disturbingly, clear that this "problem" (that internal evidence is compatible with the falsity of what is claimed to be known) applies generally to inductive inferences. There has to be something more specifically wrong with criterial evidence if McDowell is to have any chance of worrying us about it.

McDowell also characterises the "problematic" feature in a more problematic way, where it is put as the fact that the criterialist's claim to know that someone is in an 'inner' state is accompanied by this feature: for all the criterialist knows, the person may not be in that 'inner' state. Though he accepts that the incoherent position rejected earlier is indeed not what is in question, he writes that we are faced nevertheless with

> ... the tempting thought that, since 'criteria' are defeasible, someone who experiences the satisfaction of 'criteria' for the ascription of an 'inner' state to another person is thereby in a position in which, for all he knows, the person may not be in that 'inner' state. (p. 458)

Now this is obviously not to be construed as repeating the characterisation that was found to be incoherent. So we need to understand it as the claim that for all that one *otherwise* knows, the person may not be in the 'inner' state. We are, that is to say, to leave aside whether we know or not that they are in such a state, and note that for all the criterialist otherwise knows, they may not be in that state. This is then admirably consistent with the other characterisations (quoted above), but unhappily shares their property of seeming straightforwardly to apply equally to inductive inferences in general.

This makes the more pressing the intriguing question of what support is offered for the claim that the feature in question is problematic. The target is ridiculously wide and so far no support has emerged. However, McDowell's claim is, indisputably, in fact, limited (at least in intention) to criterial evidence.

> How can an appeal to 'convention' somehow drive a wedge between accepting that everything that one has is compatible with things not being so, on the one hand, and admitting that one does not know that things are so, on the other? (p. 458)

He is at pains to point out that he is not attacking inductive inferences in general.

> This line of thought may seem to be an indiscriminate attack on the idea that knowledge can be based on an experiential intake that falls short of the fact known (in the sense I explained: namely, being compatible with there being no such fact). That would put the line of thought in doubt; but the objection fails. We can countenance cases of knowledge in which the knower's epistemic standing is owed not just to an experiential intake that falls short of the fact known, in that sense, but partly to his possession of theoretical knowledge: something we can picture as extending his cognitive reach beyond the restricted range of mere experience, so that the hostile line of thought does not get started. But that cannot be how it is in the 'criterial' cases. To hold that theory contributes to the epistemic standing, with respect to a claim, of someone who experiences the satisfaction of 'criteria' for it would conflict with the insistence that 'criteria' and claim are related by 'grammar'; it would obliterate the distinction between 'criteria' and symptoms. (p. 459)

The thought is surely that those cases, more generally, where there is a gap between what one otherwise knows and what one claims to know, are acceptable provided there is a bridge, something enabling one to cross the gap. For the gap to be crossed the thought is that we need theoretical knowledge. That is, of course, a vague notion, but I think it is plain enough that McDowell would see the analogical inference to other minds as possibly qualifying (almost certainly only possibly). He would see it as precisely distinguished from his target because it appeals to correlations and not criterial connections. That much has to be accepted on pain of depriving McDowell's attack of any sense whatever

So McDowell's attack is an attack targeted specifically, and explicitly, on the notion of criterial evidence. However, it is not defeasibility as such that is the problem. It is the twin features of defeasibility and the conventional nature of criteria that together are claimed to create the problem. So inductive evidence in general is not the sort of evidence threatened by McDowell's attack, which is unsurprising, since the whole point of criterial evidence is to be non-inductive. However, so far we seem to have been provided by this attack with a distinguishing feature that separates criterial evidence from inductive evidence rather than an argument that criteria, being conventional, cannot, given that fact, survive defeasibility. At best, it has been shown that criterial evidence cannot bridge the gap in the way inductive inference may bridge the gap but that does not show that criteria cannot bridge the gap (of course, this chapter has been providing arguments as yet unprovided by McDowell, that they cannot).

McDowell asks us to consider a pair of cases (and in doing so, seems to write as if an argument is being provided)

> in both of which someone competent in the use of some claim experiences the satisfaction of (undefeated) 'criteria' for it, but in only one of which the claim is true. According to the suggestion we are considering, the subject in the latter case knows that things are as the claim would represent them as being; the subject in the former case does not. (In both cases it would be 'criterially' legitimate to attribute the knowledge, but that is not to the present purpose.) However, the story is that the scope of experience is the same in each case: the fact itself is outside the reach of experience. And experience is the only mode of cognition - the only mode of acquisition of epistemic standing - that is operative; appeal to theory is excluded, as we have just seen. So why should we not conclude that the cognitive achievements of the two subjects match? How can a difference in respect of something conceived as cognitively inaccessible to both subjects, so far as the relevant mode of cognition goes, make it the case that one of them knows how things are in that inaccessible region while the other does not - rather than

> leaving them both, strictly speaking, ignorant on the matter? (pp. 490-60)

But there is nothing here that does anything other than repeat the claim that theoretical knowledge is excluded in the case of criteria. If we remove that claim we could rerun the passage as an argument against inductive inference in general.

McDowell keeps hammering the fact that it is the conventionality of criteria that, given defeasibility, does the damage. There is a gap which has to be bridged. Convention cannot close that gap, cannot decree an entailment from behaviour to inner state (Behaviourism did not appeal to convention). I think McDowell's "argument" is precisely that convention, given the nature of convention, cannot bridge the gap opened by defeasibility. One cannot decree, as it were, that the gap is crossed. One cannot stipulate that it is crossed (by criteria, or whatever). The bridge has to be constructed from real materials, not from such stuff as dreams are made on. Canute ordered the tide to stop. Canute had the advantage over the criterialists that the tide manifestly disobeyed. They can keep their illusion that they have succeeded, uncontradicted by experience. But the fact remains that a mere convention cannot do what has to be done here. In effect, McDowell's complaint is that criterial "evidence" is "evidence" in a game. It is essentially make-believe evidence. The gap cannot be closed by claiming that we just do accept that certain behaviour is adequate evidence. McDowell perhaps thinks these points are so obvious they do not need to be made.

Paul Robinson has attacked McDowell's attack on criterial knowledge.[28] Robinson rightly presses hard on the danger to McDowell's attack of its being equally an attack on inductive inference in general. He argues that McDowell is, in fact, committed to the denial of the claim that

> We can know that p even though one's basis for believing that p is compatible with not-p[29]

We have seen that McDowell does not take himself to be committed to such a denial. Robinson thinks he cannot evade the charge. I side with McDowell against Robinson on this matter. Until the end of his examination of McDowell's position, Robinson takes it that defeasibility is the problem, according to McDowell. But that is manifestly not the case. It is defeasibility attached to the conventionality of criterial evidence that works the mischief for that sort of evidence. My objection to

McDowell is that he can thereby distinguish criterial evidence from inductive evidence but that there seems to be no argument for his hostile claim against criterial evidence other than that the criterialist cannot appeal to theoretical knowledge to bridge the gap between the accessible and the inaccessible. That criteria are conventional is crucial to that differentiation which, nevertheless, does not constitute an argument that criteria cannot otherwise bridge that gap.

Robinson, right at the end of his article, does consider the suggestion that it is the conventionality of criteria that makes the difficulty for criterial knowledge. He comes to that thought by way of considering the point that if 'facts lying beyond our cognitive reach can be known to hold' then criterial knowledge must be suspect other than on the ground of its defeasibility:

> if there is a problem [it] is not merely that it rests upon a defeasible basis.[30]

Unfortunately, Robinson forgets to note that the idea is that it is not *merely* defeasibility that is the problem. What he considers is the suggestion that it is merely conventionality that is the problem .

He quotes McDowell:

> How can an appeal to 'convention' ... drive a wedge between accepting that everything that one has is compatible with things not being so, on the one hand, and admitting that one does not know the things are so, on the other? As far as its bearing on epistemological issues is concerned, the 'criterial' view looks no more impressive than any other instance of a genre of responses to scepticism to which it seems to belong: a genre in which it is conceded that the sceptic's complaints are substantially correct, but we are supposedly saved from having to draw the sceptic's conclusions by the fact that it is *not done* - in violation of a 'convention' - to talk that way. (p. 74)

His response is to invoke Sidney Shoemaker's work on criterial evidence where

> criterial evidence is represented as conventional in the linguistic sense that it is part of the meaning of, for example, a term like 'pain' that certain kinds of behaviour count as defeasible evidence for its

> ascription. Now when we infer from the fact that the animal before us is a vixen, to the fact that the animal is female, we are relying on a convention of precisely this linguistic kind. (Of course vixenhood provides *indefeasible* evidence for the claim that an animal is female, but that is beside the present point.) Yet there is no temptation here to question the ability of that convention to deliver knowledge that the animal is female. So it looks as though it cannot be a *general* feature of conventional evidential links between claims that those links are epistemologically suspect. The task McDowell now faces, therefore, is to explain why it is that defeasible evidence, in particular, which is conventional in this sense, is less serviceable epistemologically speaking than non-conventional, or, as Wittgenstein sometimes says, 'symptomatic' evidence. Until we are told exactly how this task might be completed, we have, I think, no compelling reason to share McDowell's mistrust of conventional evidential relations. (p. 75)

This will not do. I agree with Robinson that McDowell should give us an argument which supports the claim that conventionality cannot cope with defeasibility. But the fact that indefeasible conventional criteria can provide evidence is irrelevant to that issue. McDowell's response ought to be that precisely the fact that indefeasible criterial evidence provides knowledge serves only to underline the problem for defeasible criterial evidence. We know in the case of indefeasible criterial evidence whence the knowledge comes. That someone is a bachelor entails they are single. There is no such entailment to appeal to in the case of defeasible criterial evidence. The entailments available are those I have examined already in this chapter, to their complete discredit. That someone is single, neither entails that they are a bachelor nor, despite the conceptual connection, is any significant evidential use, if what we want to know is whether or not that single person is a bachelor. That sort of inference, I am sure McDowell would agree, is McDowell's target.

So it is not claimed by McDowell, or anyone else, that 'conventional evidential links' are generally suspect. The claim is that conventional, defeasible, evidential links are unacceptable. I have claimed he has no explicit argument for this claim other than the point that criterial evidence cannot fall back on an appeal to theoretical knowledge since criterial evidence is, precisely, not empirical. I have insisted that fact does not constitute an argument. It serves rather to point out that for the criterialist one way out is blocked. I have attempted to extract an argument that I think is implicit in McDowell. That seems to me plausibly what he would endorse. It seems independently of that, convincing. In McDowell's defence, it could be claimed that just as he owes us an explicit argument, so do the criterialists. Criteria cannot be put forward as a magical solution. That criterial evidence is not empirical is true. That it is evidence needs support. Though McDowell has failed to comply with his obligation, the explicit arguments I have examined from the criterialist side have turned out to be bad. They could be seen as attempts to bridge the gap other than by the use of theoretical (inductive) knowledge. They have been shown to be unsuccessful.

CHAPTER SIX

THE "OLD" PRIVATE LANGUAGE ARGUMENT AND OTHER MINDS

A private language, in the context of other minds, is a language that is logically impossible for anyone other than its user to understand. It is claimed that such a "language" is impossible. What has this to do with the problem of other minds?

Let me approach this by way of considering Robinson Crusoe, Antonio Udina (the last speaker of Dalmatian), and you, writing in your very own code in your diary. These are not cases of a private language in the requisite sense since they are not logically impossible for others to understand. However, all three speak, and write, on their own.

I shall call those who claim that a "language" which is logically impossible for others to understand is not itself possible, Privateers. Privateers claim such a "language" is impossible because a language requires that it be possible that a user's practices be checked up on by others. This possibility is needed because a language needs the distinction between correct and incorrect use and this depends on the possibility of others being able to check up on the use made of the language. Where it is impossible for others to check, whatever *seems* a correct use will *be* a correct use, which means that the distinction has disappeared. So at any rate runs the Privateer's line of thought.

Robinson Crusoe is fine. Despite his being on his own, and, therefore, unable to appeal to others, what seems to him a right use might not be a right use and, in principle, others could check on his use. What seems to him to be right is what he is stuck with but the distinction remains, in that others *could* distinguish between correct and incorrect use. The distinction can be made, and made by him, in terms of possible checking. The user of the necessarily private "language" cannot make the distinction, so Privateer claims, because, in principle, they cannot escape from how it seems to them. There is nothing else.

The point is not that if others can check then so can Robinson Crusoe, on his own. It is that if others cannot, then Robinson Krusoe cannot.[1] The distinction between correct and incorrect use disappears in the latter case. That is why Robinson Crusoe is fine and Robinson Krusoe is not. So what seems the reasonable view: Privateer thinks that if others can check so can the solitary user and that is what makes Robinson Crusoe acceptable, is not right. If others cannot check, neither can the solitary. If others could check, in principle, the distinction between correct use and incorrect use could be made, in principle, however it seems, to Robinson Crusoe both generally and in particular cases. The nature of language is what enables the distinction to be made, not the situation of the user.

Let us return to the problem of other minds. It follows from Privateer's position, that the Private Language Argument is not an argument to other minds in any way whatever. There need not be anyone in existence other than its solitary user for a language to be meaningful. So we cannot argue from the meaningful use of a language to there being other users. So the need for argument to other minds remains. Provided a language allows the distinction between correct and incorrect use it escapes Privateer. It is only threatened by Privateer if it is in principle impossible for anyone other than its user to understand it. Arguments to other minds as such are acceptable, even analogical inferences from one's own case are acceptable, provided they avoid offending Privateer where he is sensitive.

So the analogical inference to other minds so far escapes Privateer. However, Privateer does come into conflict with what is often a central belief of those who run the traditional analogical inference to other minds. The analogical inference to other minds is not in trouble on the ground that the analogical arguer cannot appeal to others without begging the question. Robinson Crusoe cannot appeal to others. It is true that it is logically impossible for the analogical arguer to appeal to others without begging the question. But it is logically impossible for Robinson Crusoe *qua* Robinson Crusoe to appeal to others. Privateer is not interested in these logical impossibilities. Privateer's quarrel is with a "language" such that only its user can understand it. The logical impossibilities just listed are irrelevant to that. They do not result in its being logically impossible for others to note that incorrect use is occurring. Until that happens, the distinction between correct and incorrect use can be made, and Privateer is happy.

The analogical arguer to other minds needs to fear Privateer only if they espouse a "language" such that others cannot check up on it. But if they do espouse such a language, and they generally do, then Privateer's assault is relevant.

They generally do espouse this impossible "language" because they claim we learn only from our own case about mental states. They generally claim that we know directly that we are in pain and we know only from our own case what pain *is*. So others cannot, in principle, check up on our use of 'pain'. Others lack the capacity to have the direct knowledge required. That is Privateer's point of impact on the traditional analogical inference to other minds.

So arguments to other minds, and analogical arguments from one's own case to other minds are in order for Privateer provided they use an acceptable language. That is what is at issue, not that the analogical inference is unacceptable while, say, a scientific inference is acceptable. The issue is whether or not the analogical arguer to other minds insists that the use of mental terms depends on the user's direct knowledge which is, consequently, uncheckable by others. If they avoid that, they avoid Privateer.

Given the nature of Privateer's assault, it cannot be repulsed by the claim that if checking occurs, or were to occur, the results would have to be filtered through the user. In the case of the acceptable language, none of us can escape from how it seems to us in that if we have been checked up on we depend on its seeming to us that we have indeed been checked up on. We cannot escape in that sense from how it seems to us.

This is true but irrelevant. Robinson Crusoe teaches us that the possibility of checking is the issue. In the offending "language" the user cannot escape from how it seems to them in the sense that it is not possible in such a "language" to have a clash between how it seems to them and how it seems to another. That one cannot escape from how things seem to oneself (that one clashes or does not clash) is not the point. So at any rate runs Privateer's argument.

At this stage Privateer's position needs to be modified for their own good. The "language" such that only one person can, in principle, understand it, is claimed to be impossible. But what follows directly from the fact that only one person can understand it is not that evidence of a breakdown in correct use is unobtainable, but that it is unobtainable for, and from, others. Given that the problem, according to Privateer, is that the distinction between correct and incorrect use cannot be made in this "language", then precisely in those terms is how the impossible language should be characterised. It is impossible to have a language where no distinction can be made between correct and incorrect use. That is Privateer's central claim. The traditional "language" of the mind is one such, according to Privateer.

Thus modifying Privateer we avoid irrelevant musings about what might make a language logically impossible for others to understand or even whether it is logically possible to have such a language. We also avoid the misconception involved in responding to Privateer by insisting that the analogical arguer to other minds *can* be understood by others. As far as Privateer is concerned, the targeted analogical arguer, professing to learn what pain is from their own case, does not have a language of "pain". So others *cannot* understand such an arguer, since there is nothing to understand.

I have claimed that the Private Language Argument is in no way an argument to other minds, nor does it outlaw arguments to other minds. If it were thought that there would be no problem of other minds once we avoided offending Privateer, then the argument of Chapter Two would in any case dispel that thought, for it was argued there that no theory of mind avoids the other minds problem or makes it easy to solve. So an account of the mental which avoids Privateer's assault will not avoid the other minds problem. However, independently of that argument, Functionalism, for example, ought not to offend Privateer; at any rate it does not offend *against* Privateer, but it has the other minds problem.

It is worth getting quite clear about why the other minds problem survives Privateer. The other minds problem is generated by any claim that direct knowledge is available in our own case but not in the case of others. It is this asymmetry that generates the problem. The asymmetry is consistent with indirect knowledge of other minds being available, via some argument or other, which is epistemically as potent as direct knowledge, providing complete certainty.

Privateer's problem is generated by any claim that knowing the nature of mental states *depends* on one's own case. The two claims are distinct. The claim that offends Privateer entails the claim that generates the other minds problem. But the claim that generates the other minds problem does not entail the claim that offends Privateer. That is why Functionalism, for example, does not fall foul of Privateer but does generate the other minds problem, for direct knowledge of one's own mental states applies for Functionalism. That is independent of the additional argument (in Chapter Two) that Functionalists need to use an inference based on their own case. That inference does not as such offend Privateer and it does not in the Functionalist case because there is no suggestion that the functionalist language cannot be checked.

Another line of thought that, given a "public" language, there would be no problem of other minds, or there would be, somehow, an inbuilt solution, is the following, perhaps to be associated with Norman Malcolm.[2] If a language is checkable by others then there must be a way of knowing that the user of that language is using it correctly or incorrectly. If so, surely the other minds problem is solved.

This line of thought has several problems. One is that correct and incorrect use of a language might occur by other than creatures with mental lives and, presumably, checking might occur by such creatures. So language use and its checkability is not, as such, connected with minds. Another problem is that the language has to be checkable in principle. It has to be *possible* that it be checked by others. This has to be understood as: were there to be others they could check. Whether there are, in fact, others (checkers) is a separate issue. *How* it might be established that there are, in fact, others is a separate issue. Privateer's insistence is that the language passes this test: if there were others they could check on its correct use. At best, we have a guarantee from this line of thought that if there is a (genuine) language then there must be a way of checking on its correct/incorrect use, so there must be a way of establishing that we are confronted with a language user thus (if we ignore the previous problem) with an other mind. But this test provides us with no information about what that way might be. Perhaps it is the way through criteria, perhaps it is the analogical inference to other minds, or whatever. The only constraint is that the approach to others minds is not to offend Privateer.

A final line of thought, before I look at whether there is anything to be said for Privateer's position, is that if the language is "public" and checkable then mental terms must refer to something checkable, therefore public. The mind is no longer "hidden". What is (blatantly) wrong with this is that the public language is not to be public in the sense that what its terms refer to are required to be publicly observable. That meanings are to be public does not mean that what a language user refers to is to be observable. It means only that what is meant by a term in the language is subject (in principle) to being checked by more than one person. Again, the Functionalist account of pain meets the checkability condition, but, that someone is in pain on the Functionalist view is directly observable by (at most) only one person.

I have argued that Privateer's target is the notion that the meaning of terms in a language could only be learned from one's own case. If we remove the reference to how the meaning is learned, the target notion is that the meaning of terms might be known only through direct knowledge in one's own case. I have argued that even if Privateer succeeded, arguments to other minds (including analogical inferences from one's own case) would as such still be in order and would, further, be needed. Such arguments are safe from Privateer provided they avoid the targeted notion. However, do these arguments have to avoid the targeted notion? Does Privateer's attack on the notion succeed?

The purest case to consider is the "E-type" sensation made famous by Wittgenstein. This is hypothesised to occur unaccompanied by behaviour and with no correlation whatever to particular circumstances. That is what makes it pure.

> Let us imagine the following case. I want to keep a dairy about the recurrence of a certain sensation. To this end I associate it with the sign "E" and write this sign in a calendar for every day on which I have the sensation. - I will remark first of all that a definition of the sign cannot be formulated. - But still I can give myself a kind of ostensive definition. - How? Can I point to the sensation? Not in the ordinary sense. But I speak, or write the sign down, and at the same time I concentrate my attention on the sensation - and so, as it were, point to it inwardly. - But what is this ceremony for? (sic) for that is all it seems to be! A definition surely serves to establish the meaning of a sign. - Well, that is done precisely by the concentrating of my attention; for in this way I impress on myself the connexion between the sign and the sensation. - But "I impress it on myself" can only mean: this process brings it about that I remember the connexion *right* in the future. But in the present case I have no criterion of correctness. One would like to say: whatever is going to seem right to me is right. And that only means that here we can't talk about 'right'.[3]

The purity of the "E-type" sensation does not make its situation irrelevant to other "impure" sensations occurring in specific circumstances and accompanied by characteristic behaviour. The "E-type" sensation is relevant to the meaning of any sensation term provided how the relevant sensation feels is at least part of the meaning of the sensation term. Privateer would reject that possibility and that possibility would have to be rejected if the "E-type" sensation is rejected.

The classic attack on the "E-type" sensation is thus that no distinction can be drawn between correctly following the relevant rule and failing to follow it, between correct and incorrect use of 'E'; for whatever is going to seem right will *be* right. Now this is simply not so. Getting it right will be a matter of using 'E' consistently, in accordance with my conception of the "E-type" sensation. If I do not, then I fail to use it correctly, however things seem to me. In particular, if I apply 'E' in all sincerity, to sensations variously other than E, then I fail to use it correctly on any further occasion of its use.

What would *show* that I am right? Nothing will do this, just as nothing will *show* that I am wrong. But it does not follow I am not entitled to believe I am right and entitled to be sure about that. Why should I be doubtful? I need reason to be doubtful and mere possibility is not a reason. The possibility that my use is inconsistent is balanced by the possibility that it is not. I am entitled to believe I am consistent until provided with a reason for being doubtful. So the distinction between correct and incorrect use remains even if it will not show up either way to the user.

Why should the possibility of an appeal to others make a difference? Were others involved we should then have something more to go on than how it seems to the user but this would not, ultimately, provide any escape from how it seems to the user. Appeal to others, correction by others, corroboration by others, is condemned for the user to be always nothing more than this: as it seems to the user, they appeal to others, are corrected by others.

But the mere possibility of appealing to others does nothing. Were it to be true that it could be clear in the public domain that a user is not using the term 'wattle' correctly (ignoring the fact that perhaps they are but their perceptual system is awry) the *possibility* of this does not provide evidence that the user is or is not using the term correctly. The possibility is no epistemic use.

That leaves us with the bald claim that its absence means there is no

> ... difference between 'He is following a rule' and 'He is under the impression he is following a rule' ...[4]

The role of the possibility is to provide a distinction otherwise claimed to be unavailable. But this is a mistake. It does not follow from the fact that a user could not discover that they were in error that they are not. It does not follow we cannot distinguish (make the distinction) between correct and incorrect use because we cannot distinguish (differentiate) between such cases in practice. The user is entitled to trust their use of the "E-type" term is correct until provided with a reason for being doubtful. The same applies to the word 'wattle'.

So, (i) the possibility of appeal to others has no epistemic role; (ii) the possibility of appeal to others is not needed for there to be the needed distinction between correct and incorrect use, and (iii) actually appealing to others is no magical deliverance for the user from dependence on how it seems to the user.

What view of sensations is acceptable to Privateer? The short answer seems to be none. If knowledge of them depends on one's own case then according to Privateer, no one, including their owner, has any knowledge of them. If we have any picture it is of Wittgenstein's "beetle in the box".[5]

> If I say of myself that it is only from my own case that I know what the word "pain" means - must I not say the same of other people too? And how can I generalize the *one* case so irresponsibly?
>
> Now someone tells me that *he* knows what pain is only from his own case. - Suppose everyone had a box with something in it: we call it a "beetle". No one can look into anyone else's box, and everyone says he knows what a beetle is only by looking at *his* beetle. - Here it would be quite possible for everyone to have something different in his box. One might even imagine such a thing constantly changing. - But suppose the word "beetle" had a use in these people's language? - If so it would not be used as the name of a thing. The thing in the box has no place in the language-game at all; not even as a *something*: for the box might even be empty. - No, one can 'divide through' by the thing in the box; it cancels out, whatever it is. That is to say: if we construe the grammar of the expression of sensation on the model of 'object and name' the object drops out of consideration as irrelevant.

Before I draw out the implications of this fable let me point out that the supposed lesson of the "E-type" sensation is that there could not be a name for these beetles. Provided they are such that it is not possible for anyone to see any beetle other than their own then if on successive viewings it seems to the owner to be that beetle again (still) then too bad. Given no one else can check, the distinction between its being the same or different is claimed to disappear. But why should it not be the same beetle? It will or will not be the same whatever the owner thinks. The "E-type" reasoning is particularly implausible in the beetle case, though Wittgenstein does not invoke it and is making a distinct point.

The beetle case is certainly a model for sensations on the traditional view of sensations. On that view everybody's "pain" might be different, might be absent (in the case of others). But, of course, equally, everybody's "pain" might be the same and not absent. The various arguments to other minds seek to support the latter claim.

However, the immediate point to note is that for Privateer, sensations, like these "beetles", lack any features. They are not to be found outside their "boxes". Nothing can be likened to them. Nothing can be said about them. Nothing can be known about them, even by their owner. These "beetles"/"sensations" are truly mysterious. So much so, they get rid of themselves, which saves us the trouble. There might well be nothing there. There might as well be nothing there.

The upshot is that Privateerism commits Privateer to Pure Functionalism, if Privateer were to be in the business of adopting a theory of mind. It is the only conceptual account of the mind otherwise defensible that they could judge respectable. Consider the possibilities in the simplified case where pain is taken to be correlated simply with wincing (pain without correlations is clearly to be ruled out). One possibility for Privateer, as an elucidation of our concept of pain, is Behaviourism. Another is to regard pain as: that which, whatever it be, causes wincing (where it does not matter what occupies that role). The third possibility is to regard pain as: what typically causes wincing. But Privateer cannot accept this third possibility since it involves differentiating between what typically causes wincing, and what else might cause wincing. It allows that what typically causes wincing might not cause wincing and might just occur by itself. The third possibility is committed to pain's having internal features which are distinguishable from other internal features.

Privateer can choose only between Behaviourism and Pure Functionalism. Mental states can be banished (the vanished "beetle") or can be internal states with a certain role and otherwise featureless (it does not matter what a "beetle" is provided it is in the box and "beetles" cannot exist outside boxes). As soon as internal states are allowed to have features (qualia) other than connections with behaviour and circumstances then Privateer must reject them. Privateer must do so since such features, being internal, can play no role in the concept of pain.

I am taking it that Behaviourism will be rejected. I am not claiming that Privateer, even if in the business of selecting a theory, must plump for Functionalist Materialism. That is a different issue. However, how they would fend off Functionalist Materialism is not clear. To fend it off surely requires invoking features of the mental that resist an account of them as purely material. Such features will have to be internal. How then could Privateer invoke them? I am not, however, insisting that Privateer's embracing of either Behaviourism or Pure Functionalism shows that Privateer is wrong. My point is that Privateer is condemned to that choice. There is no other option open to Privateer.

I have insisted that the central issue for Privateer is whether meaning could exist without the possibility of checking. Otherwise Robinson Crusoe would be in trouble and Antonio Udina and our private (coded) diarist. what difference would it make it we switched to actual checking? The requirement now becomes that it is a condition on meaningful use of language that the user has been checked. Perhaps this will allow headway on the other minds problem?

Such checking would have to be by other people if headway was to be made. But how can it be established that such checking has occurred without dealing with the other minds problem? We could only avoid this by insisting that each of us knows that we use language meaningfully. But how is this known? Perhaps we only *think* we do.

It should be noted that this route to other minds, in order to give us other minds, gives us other minds exactly like our own since the checkers have to be like us. That would be because the idea is that we should have been instructed in our language by others who are like us. The various terms must all mean the same. If 'pain' has a public meaning then my pain is just as yours is. Ayer's[6] child raised by robots should despatch this line of argument. The child learns the language at the "hands" of robots. The child's pain is likely to be the only pain in that community.

What is wrong with this whole line of argument, apart from its being question-begging, is that the demand for checking cannot be a demand for checking by those exactly like us. It is irrelevant who, or what, does the checking. All that can be demanded is that the use of the language be checked (monitored) so that correct use is the outcome. If the language is public then the checkers can be other than human, can be machines. Other minds cannot be reached this way.

The more homespun version of this "argument" for other minds is that we are social creatures and language is a social instrument. This invites, I think, a "ho-hum" response until the attempt is made to use it, say, to give us other minds. When pressed in this way, it is clearly question-begging and goes the way of the less homespun version.

CHAPTER SEVEN

THE "NEW" PRIVATE LANGUAGE ARGUMENT: KRIPSTEIN'S* NEW FORM OF SCEPTICISM

> Wittgenstein has invented a new form of scepticism. Personally I am inclined to regard it as the most radical and original sceptical problem that philosophy has seen to date, one that only a highly unusual cast of mind could have produced. Of course he does not wish to leave us with his problem, but to solve it: the sceptical conclusion is insane and intolerable. It is his solution, I will argue, that contains the argument against 'private language'; for allegedly, the solution will not admit such a language. But it is important to see that his achievement in posing this problem stands on its own, independently of the value of his own solution of it and the resultant argument against private language.[1]

What is this new form of scepticism? It is that a single language-user, as an isolated unit, cannot be said to mean anything by any of their utterances. Meaningful utterance necessarily depends on shared responses. Content can be given to the notion of an individual's utterances being meaningful only by reference to others. What is new in Kripstein as compared with the "old" Private Language Argument, is an insistence that there is no content to meaning something by an utterance other than uttering words and having other language users agree that the utterance is appropriate. In effect, it is to make the move to the Pure Functionalism that I argued (in Chapter 6) was the only decent refuge for Privateer, but, having made that move, to add the insistence that there is no content to meaning other than shared responses.

This radical position is supported by way of arguing that the alternative views fail. Any alternative would have to distinguish between an individual's utterance meaning one thing rather than another, or, nothing at all; between an individual's seeming to follow a rule when they speak and really following that rule, as opposed to following some other rule or none at all. Additionally, it would have to *justify* what the individual does. The fact that others (would) agree, or not, succeeds in meeting these various requirements. The alternatives fail. So, at any rate, Kripstein argues.

The broad alternative to Kripstein is clear enough. What is meant by a speaker's utterance is determined by the rule being followed. What is meant is not determined by a single rule which covers previous uses of the term because there will always be more than one rule that does that. But the rule followed in the past will determine the application in the new case. That will be why speakers do what they do. It will not be arbitrary. It will have been settled in advance of their making the new application. Given what they have been doing, what they do now is what they have to do if they are to go on in the appropriate way.

Kripstein's view, by contrast, is that how to go on in the new case is arbitrary other than that others would do as we do. If they would not then insofar we are mistaken. But to be right is simply to have others agree. Until the choice is made and others concur then, there is no content to the notion that one choice rather than another is right (other than what others *would* do). At all stages the only facts are what is done. In particular, one does not know that one should go on a certain way; one does not discover that one should go on a certain way; one does not recognise that one should go on a certain way; it is in no way true that one ought to go on a certain way.

Kripstein's strategy in arguing for these claims is to reject the alternatives that would allow sense to be made of following a rule, of meaning something by an utterance, other than by reference to community assent. Those accounts which refer only to facts about the individual rule-follower are argued to fail. Such accounts come down to two: accounts which refer to mental facts and accounts which refer to dispositional facts. Here is how Kripstein's attack proceeds.

The dispositional suggestion is that meaning something is not a matter involving an occurrent mental state but of one's being disposed to act in a certain way. But that one is so disposed is true but unhelpful, for this is compatible with its being arbitrary that one responds one way rather than another. Even if it is preordained it may still be arbitrary. What was true in the past, what is there now, that establishes that what one is now disposed to do is what one *was* disposed to do? If there is a fact (one's past disposition) how does one have access to it and how would it *justify* what is done now?[2]

Information about what will be done, or would have been done, does not tell us what *ought* to have been done. The fact we are after has to *justify* responses. It also has to determine that one response rather than another is the one that *ought* to be made. One might not make the response one ought to make or one might make it accidentally, so the dispositional account will not do.[3]

What of mental facts? The problem here is how any mental state could do the trick. It has to determine how new cases are to be handled. With arithmetical calculations it is clear that no image or qualitative state will suffice for this. But it is clear enough also for other, more plausible cases (e.g. colours, shapes).[4] In any event, it does not seem there *is* an introspectible meaning experience. Even if there were, how would it help with the new cases?

Replacing talk of images or qualitative states with talk of ideas, notions, or internalisations, does not help. How would any such state account for our giving the right answer? How might the (infinite) new responses be contained in it? We could just insist they are but that seems to be whistling in the dark. Some account is needed of how the trick is managed. That we are in the state guarantees that we mean one thing rather than another, determines how we deal with new cases, and justifies our responses. It seems wholly mysterious how it does all this, so saying it just *does*, seems empty.[5]

It should be noted that vagueness is not the issue. There is no vagueness in the rules of arithmetic. In the case of, say, something's being green, Kripstein's sceptical problem may be raised about the central cases. Indeed, the case of borderline greens is instructive. Here one expects hesitation, uncertainty. Were it that the community just did agree, *after* hesitating, we should find that interesting and would note the contrast with the standard cases. In actual fact, the borderline is an area of persisting hesitation and disagreement. Such hesitation and disagreement are distinct from what is involved in Kripstein's problem.

Nor is that sceptical problem one that involves the notion of family resemblance. (1) A family resemblance concept need not be vague. (2) Where it is vague the point just made about vagueness applies. (3) It might be thought relevant that a crucial feature of family resemblance concepts is the judgement that one feature resembles another and that this is bedrock in that all we can do is find it similar or not. But this is a controversial view of resembling and, in any case, Kripstein does not invoke family resemblance in any relevant way.

Nor is ostensive definition the problem. The sceptical point may be presented in terms of ostensive definition, but the various hazards of ostensive definition are to be distinguished from the basic fact, that there is no fact to be understood, grasped, recognised.

It should be made clear also that the difficulty is not dependent on accepting narrow behaviourist or verificationist criteria. An element in Kripstein's attack is an invitation to introspection and the claim that introspection will not produce what is needed. As Kripke frequently points out, even God would fail to turn up what would have to be turned up.

The relevance of all this to the Private Language Argument is indirect. The relevance is not that such a language (one that it is logically impossible for others to understand) becomes impossible given that it is limited to facts about the individual utterer. The relevance is in the "solution" to the sceptical problem. That solution insists that sense can be made (but can only be made) of rule-following by bringing in a group of utterers (at any rate, we need more than one). The crucial notion is of agreement about utterances between utterers. A language such that necessarily only the utterer understands it cannot be "rescued" from the sceptical paradox.[6]

Kripstein's "solution" to the sceptical problem depends on community agreement. What is true of the individual alone achieves nothing by itself. We cannot make sense of an individual's following a rule if we are limited to facts about the individual, but we can do so by bringing in facts about a group of individuals. A person can be said to be following a rule if what they do is what other people do (or would do). They cannot be said to be following a rule if what they do fails this test. To follow the addition rule is nothing more (or less) than to do what the community does (or would do) and go on in new cases as the community goes (or would go).

There are various elements to the "solution". The crucial move is to abandon truth-conditions in favour of assertability-conditions. When is it warranted to make claims that a person is following a rule? Such a warrant involves a community. On our own each of us acts (generally) unhesitatingly and without justification in making our responses. We (generally) act with complete confidence. But this fails, given Kripstein's argument, to give us the needed distinction between correct and incorrect rule-following. For that we need a community. Here it is crucial that (generally) we agree in our responses. Without thinking, we respond alike. This allows us to correct mistakes. We can exclude persons not conforming to our practice and we can admit persons conforming to our practice. We can make sense in this way of following a rule and failing to follow a rule.

The idea is not that if a person is following a rule they will behave in a certain way, but rather it is the idea that if the person fails to behave appropriately we will not be justified in claiming they are following the rule.[7] Of course, if the behaviour is appropriate we are justified in claiming that the rule is being followed.[8] But it is not that what everyone does is what the rule means. Nor is it that we all behave appropriately because we are following the same rule. It is rather that if everyone agrees about what is to be done no one will be justified in claiming that to be wrong.[9]

Three key concepts in the "solution" are those of agreement, of a form of life, and of criteria. Clearly, it is crucial that we (generally) agree in our responses. We can be trained to go on as others do. How we go on, and the way in which that fits with our overall activity, is our form of life. Criteria are the observable circumstances which allow others to confirm that what someone else is doing agrees with what they do. These three concepts are all importantly involved in the key move of the sceptical solution: that others can check whether a response agrees with theirs.

The following points should be noted. (i) Reality is not of our making in that we do not *choose* how to go on in new cases. We feel impelled to respond as we do.[10] (ii) We can all be "wrong", in that a community may change its mind.[11] (iii) Robinson Crusoe is fine provided that we would agree with his responses.[12] What is essential is not that there be a community but that reference be made to a community in supporting a claim that rules are being followed. (iv) It is not that the community, being numerous, is unlikely to be wrong. There is nothing to be wrong about (or right about). We are to

> reject the idea the ... community goes right or wrong in accepting a particular verdict on a particular decidable question; rather, it just goes.[13]

It should be particularly noted that the position Kripstein adopts is not that for every rule we need an external check such that others can confirm that a person is following it. In particular, he thinks there are some sensations which do not have outer criteria. This is considered to be acceptable provided that we have shown we are part of the community in those cases where external criteria *are* available for sensations. This (less stringent) requirement still allows the "New" Private Language Argument to operate.

Further, the checkability requirement does not have to hold for every application of a particular rule. Provided competence has been demonstrated for that rule in those applications where external criteria have been available, then we are "licensed" thereby to proceed simply on how it strikes us, subject to our continuing to be able to act in agreement with others whenever external criteria are available.

It is, of course, controversial whether Kripstein's position on this is the same as Wittgenstein's. The issue is whether Wittgenstein adopts the (more stringent) requirement that competence has to be demonstrated for each (sensation) rule. Kripke thinks the matter unclear (despite Malcolm's being clear that Wittgenstein does not).[14]

I now want to look at the sceptical "solution", and how it is justified, but in the reverse order. There is not much offered by way of positive justification. Essentially the claim is that there is nothing else. Community agreement gives us a distinction between correctly and incorrectly following a rule and nothing else will.

Wright seeks to soothe the scepticism of those who find it mysterious that the community has such powers, by considering how this agreement enables my application of a term to become relevantly recognitional:

> How does others' agreement with me turn my descriptive disposition into a matter of recognition of conformity with a pattern, recognition of an antecedent fact about how the communal pattern extends to the new case?[15]

He thinks the 'solicitable community of assent just does make the relevant difference',[16] but his reassurance does not extend beyond saying, in effect, that nothing else works.

> Assuredly, there is truth in the idea that it is a community of assent which supplies the essential background against which alone it makes sense to think of individuals' responses as correct or incorrect. The question is whether it can be explained why there should be any truth in this idea unless the correctness or incorrectness of individuals' responses is precisely *not* a matter of conformity or non-conformity with ratification-independent facts; for if the contrary account were correct, the role of communal assent in determining the correctness of individuals' responses would require the capacity of the *community as a whole* to recognise what conformity to antecedent patterns of use required of us. And now essentially the same problem arises again that besets the attempt to make unilateral sense of correct employment of an idiolect. What is it for a community to recognise that it here continues a pattern of application of an expression on which it previously embarked? What does it add to describe the situation in two-fold terms, of the fact of conformity to the pattern *and* the community's recognition of the fact, rather than simply saying that there is communal agreement about the case? It is unclear how we can answer. We are inclined to give new linguistic responses on which there is securable consensus the dignity of 'objective correctness'; but we have, to so speak, only our own word for it. If 'correctness' means ratification-independent there is apparent absolutely nothing which we can do to make the contrast active between the consensus description and the correct description. Of course, it may happen that the community changes its mind; and when it does so, it does not revise the judgment that the former view enjoyed consensus. But that is a fact about our procedure ...[17]

What this comes to is that the community's role could not be what it is unless the alternative view is mistaken. So that the case for the community is the falsity of ratification independence. There is no other justification offered by Wright, no further support other than that a distinction is needed and this provides it.

According to Cavell,

> We learn and teach words in certain contexts, and then we are expected, and expect others, to be able to project them into further contexts. Nothing insures that this projection will take place (in particular, not the grasping of universals nor the grasping of books of rules), just as nothing insures that we will make, and understand, the same projections. That on the whole we do is a matter of our sharing routes of interest and feeling, modes of response, senses of

> humor and of significance and of fulfilment, of what is outrageous, of what is similar to what else, what a rebuke, what forgiveness, of when an utterance is an assertion, when an appeal, when an explanation - all the whirl of organism Wittgenstein calls "forms of life". Human speech and activity, sanity and community, rest upon nothing more, but nothing less, than this. It is a vision as simple as it is difficult, and as difficult as it is (and because it is) terrifying.[18]

That is what there is, that is all there is. It will have to do.

There are those who are dissatisfied. Cameron[19] is, I think, representative of a widespread feeling of unease at Kripstein's vision. But the representative unease seems inchoate. He does not think much of the distinction between correct and incorrect rule-following being a matter of community agreement. But he confines himself to not being sure that there would then be sense

> to our talk of "getting it right" when we ... think we are following a rule.[20]

I do not think this is a problem. At worst we should treat this as a revision of our existing notion. What is imperative is that there be a distinction which is manageable. The one offered is manageable.

> Cameron's next complaint is that we need subjunctive conditionals of the form "If you were to ask how rule R applied here, you would get this answer."[21]

But he does not say why. He merely affirms that without them it cannot be the case that

> our concepts of rules and rule-following can ever have proper application.[22]

It is likely that the two points are the same and Cameron is doing nothing more than pressing a belief that we need a notion of right and wrong other than that of agreeing with others.

His next point is that

> Whenever we engage in this activity (following a rule (as we think)) we are bound to assume the truth of something that the rule-following considerations seem to deny. The investigation-independence (in however minimal a sense) that is implied by the term 'rule' is in fact one of those fundamental presuppositions of all speech, thought and reasoning which, even if they may be unwarranted or even false, lie beyond coherent denial ...[23]

This supports the view that Cameron's two previous points are his intuition that rule-following implies investigation-independence. Apart from that intuition, the new assertions depend on the claim that investigation-independence cannot be coherently denied. He does not support this and it is, at best, not obvious where the incoherence lies.

Cameron's final point is that if the majoritarians (communitarians) are right and his strictures are right

> we would need to work out what we can possibly say about speech and thought if they ... depend essentially on a presupposition that not only lacks all warrant but cannot be given a coherent sense.[24]

But if they do so depend then we are stuck with that so speech and thought would be lost to us. Presumably the more sensible point would be: what view of speech and thought is to be taken on the majoritarian account? No doubt Cameron would find whatever was offered unacceptable.

Holtzman and Leich[25] have a more reasoned dissatisfaction. They present two criticisms of the majoritarian account. The first is its 'central difficulty', the fact that

> The account takes Wittgenstein to be denying that a language which is only, so to speak, *contingently* private, is not capable of objectivity, whereas Wittgenstein's repudiation of objectivity in Section 258 is directed against a language that is supposed to be *necessarily* private.[26]

But Kripstein is not guilty of this charge. Robinson Crusoe is fine, as was noted previously:

> The falsity of the private model does not mean that a *physically isolated* individual cannot be said to follow rules; rather that an individual, *considered in isolation* ... cannot be said to do so.[27]

An individual may be said to follow a rule provided a community *could* be brought to agree with that individual's practice. A necessarily private language fails this test. A contingently private language passes it.

Holtzman and Leich fare no better with their second criticism. In considering a possible majoritarian response to their claim that a contingently private language is "outlawed", a response which stresses that 'agreement ... in judgments' is essential,[28] they draw the "unacceptable" consequence that, on the majoritarian view,

> There is no clear sense in which the majority of speakers of a given language can be said to be *wrong* in making any one of this large class of judgments (judgements such that a speaker who disagrees cannot mean the same as the majority).[29]

Their response is that there is indeed such a sense in that the majority may not mean anything at all by some word. Their support for this seems to be that Wittgenstein was committed to this possibility. So the Kripstein (majoritarian) account fails as a consistent account of Wittgenstein. Whether this is so is not my concern. Does the "unacceptable" consequence matter otherwise? It should be noted that there *is* a sense in which the majority can be wrong (as has been pointed out previously). They can change their minds. However, it is more important to remember that Kripstein's view is that if the community agrees then no one can be justified in claiming they are wrong. Given this, if there is agreement then, if no one could be justified in claiming the community is mistaken, how, equally, would anyone be justified in claiming nothing whatsoever is meant? Presumably they might change their practice and themselves decide this, but that is the most that is possible. So Kripstein seems guilty as charged: or is he?

Holtzman and Leich seem principally concerned with whether Wittgenstein allowed for majority meaninglessness. But if *their* argument is merely that Wittgenstein did allow for it, do the considerations they separately adduce show that it *is* possible?

Their crucial point is that

> expressive uses of words ... *mean* nothing, in the sense that mastery of them does not reside in mastery of referential relations.[30]

I think Kripstein would agree, *in that sense* (Kripstein certainly *could* agree). It is not generally thought that such uses are, thereby, otherwise meaningless. Even if it were, the central Kripstein claim is that rule-governed activity is such that the majoritarian account holds. It is not that all rule-governed activity is linguistically meaningful. Such distinctions, like all others, will be subject to the majority's decision. Only *they* can opt for "meaninglessness". Given that Holtzman and Leich's "meaninglessness" involves mastery of the use of words (they illustrate their claim in terms of one use of 'I') so much "meaninglessness" is handled (adroitly) by the majoritarian account. Any more radical charge of meaningfulness would seem to be question-begging, amounting to the claim that there is more to meaningfulness than Kripstein allows.

The most common response to Kripstein is to say that the notion of community cannot rescue us from the problem. This response takes more than one form. The most popular is the following. The community proceeds blindly ('just goes'). If that is so, why should not the individual just proceed blindly? Of course, the individual does proceed blindly, as one of the individuals collectively comprising the community. So what makes the crucial difference? The claimed crucial difference is that agreement is possible between individuals. This difference is then rejected on the ground that Kripstein's argument holds against the notion of agreement.[31]

This popular response misunderstands the role of agreement in Kripstein's overall position. Kripstein's argument, if successful, works against any notion that collective agreement "creates" individual meaning over and above the facts of collective agreement. "Real" meaning is not restored to us. Collective agreement gives us, however, the needed distinction between correct and incorrect use. Agreement will be constituted by the community's agreeing that they agree. This will be a matter of its seeming to be the case that they do, making the appropriate noises. That is what constitutes agreement, as opposed to disagreement, on Kripstein's view.

It is important to keep Kripstein's problem separate from the problem of Descartes' evil demon. I might think other language users are agreeing with me when they are not. Yes I might. But that is not what Kripstein is concerned about. General sceptical anxiety is irrelevant. If the community agrees it agrees, it agrees. Its agreeing is a matter of its taking it that it agrees. There is no subterranean separate notion of (real) agreement. There is no independent fact of the matter about whether a particular response is right. There is no independent fact of the matter about whether various (of the community's) responses are right. There *is* a fact of the matter about whether the responses are in agreement. They will be if they are (collectively) said to be. Kripstein's central claim is that agreement in response is all we have and that is a matter of what we as a community take to be agreement. That is all we have by way of being right.

However, there are stickier problems for the communitarians. Blackburn has characterised their view as a

> *wooden* picture of the use of language, according to which the only fact of the matter is that in certain situations people use words, perhaps with various feelings like 'that fits', and so on. This wooden picture makes no room for the further fact that in applying or withholding a word people may be conforming to a preexisting rule. But just because of this, it seems to make no room for the idea that in using their words they are expressing judgments. Wittgenstein must have felt that publicity, the fact that others do just the same, was the magic ingredient turning the wooden picture into the full one. It is most obscure to me that it fills this role: a lot of wooden persons with propensities to make noises is just more of whatever one of them is.[32]

The crucial claims are that there is 'no room for the idea that in using their words they are expressing judgments' and that, if this is so in the individual case, then adding other people cannot help.

A dilemma can be extracted from this. The sceptical strategy, if successful, leaves no fact (other than community agreement) which will determine that the language-user means anything and no justification for what they do. There is, then, only noises. If there were more, the sceptical strategy would not work, because that more would apply in the individual case.

Cavell's characterisation of the situation (outlined previously) illustrates the problem. He writes of

> our sharing ... senses ... of what is outrageous ... what a rebuke, what forgiveness ...[33]

But on the view in question, to share a sense of what is forgiveness is nothing other than to agree on the application of the word. It is not that we ought to apply the word. We just do apply the word. What then is left of our sense of forgiveness? Surely we have no *sense* of it at all. If we have a *sense* of it, then I could have it independently.

I pointed out earlier that Kripstein does not claim that for all sensations there must be public criteria:

> the liberal version allows that once a speaker, judged by criteria for mastery of various rules, is accepted into the community, there should be some rules where there is no way for others to check his mastery, but where that mastery is simply presumed on the basis of his membership in the community.[34]

In particular:

> It is a primitive fact of our language game of sensations that, if an individual has satisfied criteria for a mastery of sensation language in general, we then respect his claim to have identified a new type of sensation, even if the sensation is correlated with nothing publicly observable.[35]

All this is completely acceptable for nothing of any substance is being licenced. All we have is words and a decision to treat certain utterings of them as "meaningful". Nothing is being said by the individual with the new sensation.

What of the community's correcting itself? This would seem to be a feature of the majoritarian account. If it is, it exhibits the emptiness of getting it right from yet another perspective. Look at this from an individual case. What if I anticipate the community's change? Until they concur I mean nothing. But if they come round to agreeing with me I shall then, retrospectively, have been right. What if I am responsible for the change (by persuading, cajoling, arguing)? Again, it remains only retrospectively that I achieve my meaning. The case seems to demand that we either accept the characterisation that noise is all there is, *or* that the majoritarian account is wrong in withholding meaningfulness until others come round.

If we consider the solitary (but persuasive) dissenter more closely we can see how that particular case is an embarrassment for the communitarians. On the traditional view majority responses are not fundamental. The majority may be corrected, its responses scrutinised and found wanting, its grasp of a concept argued to be inadequate, and its responses consequently thought to be haphazard. There is more to it, our practice would have it, than just responses. The aim is to bring others round so that they will get it right, not so that I shall then be enabled to be right.

So there must, given these communitarian woes, be something wrong with the sceptical *argument*. The dispositional account of meaning succumbs, I think, to Kripstein's attack outlined above. If so, the key move becomes the denial that any mental state could do what has to be done. Kripke concedes that it is possible to cling to this notion:

> Perhaps we may try to recoup, by arguing that meaning addition by 'plus' is a state even more *sui generis* than we have argued before. Perhaps it is simply a primitive state, not to be assimilated to sensations or headaches or any 'qualitative' states, nor to be assimilated to dispositions, but a state of a unique kind of its own.
>
> Such a move may in a sense be irrefutable, and taken in an appropriate way Wittgenstein may even accept it. But it seems desperate: it leaves the nature of this postulated primitive state - the primitive state of 'meaning addition by "plus"' - completely mysterious.[36]

Some points can be made: (1) The complete unacceptability of the majoritarian account presumably means we *are* desperate so perhaps we should prefer mystery to disaster. (2) Whatever such a state might be it is most certainly not like a sensation or other qualitative state (believing that one has a headache is distinct from having a headache). (3) It is hardly a state of a unique kind just because it is not a qualitative state nor to be assimilated to dispositions, for neither can belief, for example, it is widely thought, be so treated (neither Kripke nor Kripstein would, presumably, wish to embrace a dispositional view of belief).

Kripstein's worries take also a more specific form:

> It is not supposed to be an introspectible state, yet we supposedly are aware of it with some fair degree of certainty whenever it occurs.[37]

This is not a problem. Though it is not introspectible in the way that headaches are, that does not mean it is not introspectible. Neither is believing that something is the case. This much mystery is tolerable.

His other, 'more important' worry, is that

> it remains mysterious exactly how the existence of *any* finite past state of my mind could entail that, if I wish to accord with it, and remember the state, and do not miscalculate, I must give a determinate answer to an arbitrarily large addition problem.[38]

This is the sticking-point, as Kripstein sees it.

He has earlier asked:

> Can we conceive of a finite state which *could* not be interpreted in a quus-like [completely bizarre] way? How could that be?[39]

But if the finite state is to be a mental state and not just the voicing of words then I must be allowed to mean, for example, that it is red by uttering 'It's red'. If that state can exist then that will do what is needed. If Kripke's anxiety is taken as a genuine barrier to extrapolation then it has to be taken as denying that I can mean that it is red by uttering "It's red" or mean that it hurts by uttering "It hurts". *If* I can manage these then the other applications are determined. So that:

> There can be no such thing as meaning anything by any word[40]

is not properly a conclusion of the sceptical argument. it is built into that argument. The response to the Kripkean "Might you not mean something else?" is "No, not if I mean whatever I mean". It is only if an interpretation is taken to be wholly given in words (and other words, and so on ...) that you have the problem. If there is more to it, then I interpret the words by meaning them a certain way. And we have seen there just cannot be only the words.

I conclude that Kripstein's attack on traditional meaning fails (I take it Kripke thinks so also, having, I think, his own case against Kripstein).

And what about other minds? If the "New" Private Language Argument worked, what then? The story is the same as with the "Old" Private Language Argument. In shortened form, given that the claim is not that a community (actual) is needed but that we need to refer to a community to make sense of individual utterance, then whether or not there is, in fact, a community is a matter to explore. Put another way: there is never an issue on Kripstein's view, in or out of community, as to whether I can, within myself, without reference to a community, mean something by what I utter. I cannot do this. Whether I do or do not mean something depends on some reference being made to a community. Only such a reference could decide the issue. Whether there is such a community is a separate issue.

If that community has to be a community of minds then that would bring our problem of other minds. But agreement between language users does not, as argued in the previous chapter, demand agreement between users with minds, never mind like-minded users. The problem of other minds survives community. What is essential to the notion of community is either insufficiently "rich" to give us other minds, or, if it is sufficiently "rich" that "richness" brings with it the problem of other minds. To allow, which is in itself controversial, that the individual could easily secure "agreement" from a community, is not thereby to make available to that individual the sort of community that would give that individual companions and companionship.

CHAPTER EIGHT

"WITTGENSTEIN'S" "ATTITUDINAL APPROACH" TO OTHER MINDS

So far I have not claimed at any stage to be commenting on Wittgenstein's views, as opposed to commenting on views claimed by others to be his, or, at any rate, views claimed by others to be readily extrapolable from what he wrote. While firmly intending to remain faithful to this sensible but unheroic limitation, I want to look now at a view of Wittgenstein's thinking on other minds which seems plausible, and where that thinking interestingly seems both to intersect with and diverge from certain strands of my thinking on other minds.[1] That thinking also involves what has been referred to as the "attitudinal approach" to other minds. This remains, I think, an "approach" that is more influential than it is discussed, and it is, at least, an important historical alternative to the analogical inference. In what follows, when it is Wittgenstein's thinking on other minds as presented by M.R.M. Ter Hark that I am commenting on, I shall refer to this interpretation as "Wittgenstein".

I have characterised the other minds problem as being generated by an asymmetry, between what we know directly is the case (our own case) and what we know (if we do) only indirectly (the case of others). That is what generates the problem, which I then claim is solved by using the (hybrid) analogical inference to other minds. It is most unlikely that Wittgenstein ever thought the analogical inference was acceptable. He could I think, reasonably (if wrongly) be seen as having been initially disposed to accept behaviourism, about others, about the third person, for want of a better solution. However, he seems not ever to have been behaviourist about the first person. Indeed, his initial thinking about other minds seems to have accepted direct knowledge, acquaintance, direct observation, as holding for each of us in our own case, for the first person, so generating the problem of other minds, given the unavailability of this introspective knowledge in the case of others (the third person).

So initially there is a similar asymmetry to mine, but it is not identical if we accept that Wittgenstein was initially behaviourist about others. Such a combination, of introspective awareness of our own case, and behaviourism about others, allows a symmetry between the two cases. Direct knowledge can be allowed to be available in both cases, obtained by introspection in our own case, and by observation in the case of others. This situation remains subject , however, (as argued in Chapter Two) to the essential asymmetry generating the problem of other minds. We have direct knowledge of how it is with us. We do not have that knowledge directly in the case of others.

If we are to be neither behaviourist about others nor ourselves, then in the absence of a claim to direct knowledge of the mental states of others, we have more straightforwardly the asymmetry I have embraced. So those who deny that Wittgenstein was ever a behaviourist, would be committed to that asymmetry holding for Wittgenstein's early thoughts on other minds, where he seems to have accepted introspective knowledge in our own case (at any rate on the reading I am basing these remarks on).

This opening position on other minds (let me cautiously credit it to "Wittgenstein") is clearly unstable. The position is symmetrical in respect of direct observation, provided behaviourism was embraced for others. But that is unstable, being asymmetrical and defectively so, otherwise, in that behaviourism was applied only to others, not to oneself. This is clearly incoherent, for behaviourism is consequently to be true for me of others, including you, and true for you of others, including me. So it is to be both true and not true of me.

If it is accepted that the analogical inference is totally unacceptable, and if behaviourism (holding of ourselves and of others) is rejected, we are left with the problem generated by direct knowledge in our own case, an acceptance (no doubt tacit) of the asymmetry between that case and the case of others, and no solution in sight. This position too is unstable. Something would have to give.[2]

What "gave", in "Wittgenstein's" case, and notoriously, was the belief that each of us has direct knowledge in our own case. That belief was jettisoned. He embraced a symmetricality, insisting that knowledge did not apply in our own case. There was no introspection of, inspection of, observation of, acquaintance with "inner" states. Such a "gap" was a misunderstanding of ourselves. Each of us, in our behaviour (including verbal) gives expression to how it is with us; the "inner" expresses itself. We do not, it is claimed, know we are in pain on the basis of (introspective) observational evidence. We give expression to our pain. Our pain expresses itself through our behaviour, verbal and, of course, otherwise.

This view of ourselves is then linked to a view of others, as also expressing, in their behaviour, their pain; of their pain expressing itself in their behaviour but where our observation of their expressive behaviour is not the basis for some inferential knowledge of their "inner" self. Knowledge is ruled out, as not appropriate, in either our case or the case of others. Thus we assert a symmetricality. There is an asymmetry, in the fact that we do use observation in the case of others, but not of ourselves.

This "approach" to others spurns the analogical inference, and behaviourism. Notably though, others are, indeed, observed. Observation is clearly crucial but not in order that it might provide (observational) evidence. Others are indeed viewed, as souls, seen as souls, regarded as souls, treated as souls. The "attitudinal approach" to others is in gestation, where it is for the moment clearer what it is not, than what it is. Others are observed, as is their behaviour, but not, as I have said, in order to obtain evidence which will support an inductive inference.

However, "Wittgenstein's" focus at this early, gestating stage of the "attitudinal approach", was on the first-person, not the third-person case. What was seen as having to be abandoned in his earlier view was that each of us had direct knowledge in our own case. That claim to introspective knowledge was seen to be what was causing the trouble, generating the problem; and insolubly in the absence of direct knowledge of others (behaviourism). Given also the rejection of the analogical inference, the problem was both generated and seemed to be insoluble. What was to be done? The solution was to prevent the problem from getting started. The notion that we have direct knowledge in our own case was rejected However, the reason this notion was rejected was not so that it might be replaced with indirect knowledge, nor was it to replace the problem of other minds with the problem of our own mind, or the problem of minds, but in order to remove the problem of other minds, as a problem. Clearly, such a move (removal) had to be radical. An entire framework (subject, of course, to traditional variations) was to be replaced by a framework that would be completely different.

"Wittgenstein" and I do not completely part company at this stage, however different our frameworks. The new asymmetry is not mine but has at least one line of connection. Knowledge has been banished from the first-person case and from the case of others, but observation is needed for others, though not for oneself. Something perhaps deeper than knowledge applies in one's own case, something prelinguistic even. What is to be noted is that the thrust is to stop the other minds problem at its source. The thought was that if direct observation does not hold in my own case then the troublesome asymmetry is avoided, of direct versus indirect observation of the inner, of direct versus indirect knowledge. It is true that we observe the behaviour of others, not our own or even ourselves (through introspection). But such observation, of the behaviour of others, is not intended to provide observational evidence, which can then be the basis for an inductive (analogical) inference. Its role is within the "attitudinal approach" to other minds, an approach merely hinted at, as yet.

However, we do still have an asymmetry between ourselves and others. Observation remains crucial in the case of others. It does not apply in my own case. I need an "approach" for others, not for myself. My view of others is an attitude towards them, as souls, but I do not, symmetrically, approach myself this way. I do not have an attitude towards myself, that I have a soul. Rejecting behaviourism, and the analogical inference, "Wittgenstein" slots in the "attitudinal approach", but to others, asymmetrically, not to himself. That approach is radically different, but it surely accepts the crucial asymmetry. There is a problem about others, not about oneself. The first-person account, in terms of expression, provides the kernel of the third-person account. The behaviour of others is expressive of the inner, not to be regarded as related to the inner as observational (inductive) evidence of the inner in the traditional way. That denial (of the traditional role) is common to the "attitudinal approach" and the criterial approach. But the "attitudinal approach" (as is the criterial "approach") is to the other, not to oneself.

What exactly is the "attitudinal approach"? It allows that we are certain that others are as we are. But this certainty is not a matter of knowing that they are. It is not, of course, that we do not know they are, or that we know they are not, or that we are unsure that they are, or whatever. We are certain, but we do not *know* for certain, because our certainty does not involve knowledge. We are not certain on the basis of inferring from observational evidence. On this view, we do not know that others are as we are, nor even have a justifiable belief that they are. However, we are not sceptical, or whatever. We do not, though, merely believe that others are as we are. We have an attitude towards them, that they are as we are. But an attitude is not a belief. Whether it is prior to, or might lead to, belief, it is distinct from belief. If compatible with belief, it is independent of belief. An attitude is "deeper", it seems, than mere belief. So, at any rate, are attitudes in the context of the "attitudinal approach" to other minds. They are also, if "deeper", presumably epistemically "deeper" (not just psychologically deeper).

So any assessment of the "attitudinal approach" to other minds has to remember that what is to be appraised is not any sort of belief. What is being advanced is not that we believe that others are like us, but we just believe this, without reason, without the support of observational evidence, or any sort of evidence. It is not just a primitive fact of our lives. It is not that we just believe unthinkingly, blindly, optimistically, caringly, or whatever. Those who think, reasonably, that ignoring the problem of other minds would be better than advancing as a "solution" to that problem such unthinking beliefs, are mistaken if they think their target is the "attitudinal approach" to other minds. It is not that we just believe, without thinking, without justification, that others are like us. Perhaps the "attitudinal approach" could be seen as stressing that we do not merely believe this. What is invoked is something other, something more certain, more profound, more immediate, than anything belief, however justified, could provide. It is rather the claim that, quite simply, we do not believe in other minds. We have an attitude to them, they are seen as being like us, experienced as being like us. They are apprehended as "souls". Talk of belief misses the mark. Does each of us believe we have this "soul"? Does each of us believe that we exist? Does each of us, in pain, believe we are in pain. Talk of belief, in our case and that of others, opens a gap that is not there.

So the obvious, easy response to "Wittgenstein" on other minds will not do. It misconceives what is being suggested. Let us look more closely, with an eye on the actual object. Our attitude to others, being an attitude towards a "soul", is an attitude which incorporates an inner self. There would be no need for the "attitudinal approach" if we did not wish to embrace an inner self. There would be no need, in the absence of such an inner self, to set aside belief, with its trail of inferencing and hypothesising, and gaps between the inner and the outer. To embrace this attitude towards others is to set aside behaviourism, and the analogical inference and, indeed, all the "evidential" approaches. It is to reject the traditional gap between the outer and the inner. However, to avoid the traditional gap is not to avoid a gap. How is it thought that the "attitudinal approach" avoids a gap between inner and outer, or allows only a gap that is unproblematic?

I have pointed out that the essential asymmetry, the asymmetry that generates the other minds problem, withstands the embracing of attitudes. But there is, in any case, a clear asymmetry in the approach outlined. We do not strike up an attitude towards ourselves, as having souls. If we are not to think of our experience of ourselves as involving an inner, involving direct observation, direct knowledge, acquaintance or such like, with an inner, then we are not, even more so, to think of our access to others as involving direct observation of their inner states. But observation is relevant in their case, though not in our own. That relevance comes from the fact that our situation with respect to ourselves is different from our situation with respect to others. Why does this asymmetry not open up a gap, perhaps different, but just as much a barrier between ourselves and others as is the traditional gap?

To make headway here, we have to remember that it will not do to collapse this new approach into the old approach, back into its just being a primitive fact, a matter of psychological fact, a matter of how we are constituted, that we just have this attitude. Where it would have been claimed that we just believe that others are like us, it would now be claimed that we just have the appropriate attitude, rather than the belief. Attitude would trade places with belief. Could we not, parallel with the traditional gap, have the appropriate attitude and the "soul" be lacking? I am not saying that the problem is that it is possible that the attitude is universally misplaced (except as applied to me). That possibility is balanced by the possibility that it is not misplaced. The problem is that so far we seem to have in no way advanced beyond belief. We seem essentially to be where we were. That we just believe will not do. That we just have "the attitude" would equally not do.

At the heart of the "attitudinal approach" is the rejection of belief as the operative notion. Here is my suggestion for characterising that approach. I think what that approach amounts to is the "criterial approach" purged of belief. The role of criteria is no longer to provide justification for belief. To think they are to provide justification is to think that belief is the operative notion. Criteria, in this context, are not intended to provide (criterial) evidence, justification. Justification, being indissolubly wedded to belief, is not what criteria, in this context, are providing. So the "attitudinal approach" slips away from the strictures levelled in Chapter Five against the use of criteria in the context of other minds. Casting off the chains binding him to belief, with one bound, we might say, criterial Jack is free.

So what is the operative structure, if it is not criterial evidence? What is being relied on is a conceptual connection, a logical connection, between outer phenomena and inner phenomena. But not in order to underpin belief, and so justification. To talk of criterial evidence is totally to misconceive the strategy. Criterial evidence was to be the "middle-way" between entailment and induction. In the "attitudinal approach" we have abandoned that spectrum, abandoned that search for an alternative within that spectrum. What we have now is a different spectrum, a different (language) game, with different rules. Certainty is to be obtained from a criterial connection between the outer and the inner, but a criterial connection that is conceptual and without any links to belief. At any rate the certainty does not depend on belief. Belief is at best superfluous.

But how is certainty provided? Merely invoking the conceptual connection, even the notion of a practice, a (language) game, is clearly no different, in point of usefulness, from claiming there is a (language) game of confident belief, of belief beyond the reach of doubt, uncontaminated by justification. We would be back to the claim that we just do have this (special) "attitude" towards others.

This collapse of the "attitudinal approach" can, however, be postponed, if, perhaps, only for the moment. The essential feature, the conceptual connection, is shared with the "traditional" criterial approach. Behaviour, once again, is conceptually linked to sensation, this time because, precisely, it is expressive behaviour, expressive of the sensation. The expression (of the sensation) brings the sensation with it, as part of our conception of the expression. But we do not then invoke the notion that the expressive behaviour is thereby necessarily evidence for the presence of the sensation. Instead, our conception of the expressive behaviour is of behaviour expressive of the sensation. The behaviour is viewed as expressive of the sensation. We see it as expressing the sensation. Our immediate interpretation, understanding, is of the behaviour as expressing the sensation. That is how it is experienced. The outer comes to us stamped conceptually with the inner. We see the outer as expressive of the inner. No inference is in question, no gap between an outer and an inner that needs somehow to be bridged. As one hears sadness in the music, one sees sadness in the face, hears it in the voice. No inference is involved. There is no move from what is observed, to the sadness.

In the case of (suspected) pretending this conception is inoperative.[3] We return to belief, and inference, and evidential pondering. But that we do so fails to disturb our basic practice. On this view, such inferential pondering depends on our basic practice (of certainty inhering in the conceptual connection). We need the normal practice to render the abnormal practice possible. If we do wonder whether someone is merely pretending, we will have a reason. In this particular case we wonder about the connection between outer behaviour and the inner. But it is insisted on that we do not for a moment suspend our basic attitude to that person as someone whose expressive behaviour is generally seen as, conceived of, as inseparably connected with inner phenomena. This insistence on the normal practice underpinning the abnormal case seems independently acceptable. Pretending is something that *people* do. To worry whether someone is pretending is not to worry whether they are a person. If it were not possible in principle to pretend there would be no problem of other minds. Actual pretending, however, is a practical problem, and, indeed, the notion of pretence depends on a conception that the pretender is a person. Pretending is (at least) a mental activity. To wonder whether a human figure is pretending is to have left the other minds problem behind.

Where have we reached with the "attitudinal approach"? The certainty, the unassailable assurance, comes, can only come, from the claimed conceptual connection between the outer and the inner. Though it is possible to conceive of the (expressive) behaviour without the sensation (as in pretending), we generally do not, and we never lose our general conception of the behaviour as expressive of the inner. The certainty comes from the general inseparability, conceptually, of the expressive behaviour from its sensation.

So we do not just believe in other minds, nor just have the requisite attitude towards others. Our attitude is as it is because the behaviour of the other is seen as expressive behaviour, expressive of an inner self. That is so because we conceive of that behaviour as expressing sensation. We view the behaviour, regard it, apprehend it, understand it, conceptualise it, see it, as expressive of sensation. That is, supposedly, why we are certain. Our certainty inheres in that conception, is not obtained via some inductive argument. The immediacy of the apprehension, however, should not be confused with the claim put forward by those, like McDowell, who have, indeed, claimed that the inner self can be, sometimes, apprehended directly. However, it points in that direction, and, so pointing, the "attitudinal approach" might be seen as a middle-way between any evidential approach and a claim to have direct awareness of other selves. It would resonate, then, with the more traditional criterialist's search for a middle way between deductive and inductive evidence.[4]

The "attitudinal approach" succeeds in offering a new model for our access to other people, distinct from behaviourism, from the analogical inference, from criterial evidence, from the traditional inner/outer dichotomy. It might well describe accurately our psychological response to other human figures. However, that is as far as it goes. Nothing is provided beyond an account of why we are certain. If any sort of justification was aimed for, it is not forthcoming. The certainty is at best explained, not justified. Several examples make this clear. That people conceived of a woman as a witch did not make her a witch, however certain those who conceived of her that way were that she was. They had the relevant conception of what they observed. The model fits. What was observable, indeed observed, was conceived of as attached to what was unobservable. Here is another example, fitting the model. That men conceive of women as inferior does not (of itself) make them inferior, though men be ever so certain. Here again there seems a perfect match with the model proposed. What is in question is how what is observed is conceptualised. Women are seen as inferior, apprehended as inferior. There is no perceived gap, no inductive reasoning (at any rate, before being challenged).

It is unlikely that someone would have pretended to be a witch (perhaps some do nowadays). Such a pretence, if suspected, would not have threatened the practice of conceiving of certain women as witches. The traditional weighing of evidence would have been brought to bear on the suspect case, while the general conception remained intact. That women are conceived of as inferior does not prevent some women being judged to be exceptions. Such women do not threaten the general conception (indeed, can be seen to reinforce it). Most recently, Mrs Thatcher comes instantly to mind.

I think the "attitudinal approach" illuminates, and is illuminated by, a plausible connection with some feminist insights. They point out that women's bodies, or women as such, are generally conceived of in a certain light. What will be observable will be limited, a part only of what is involved in the conception, much of which (most of which) will be "hidden". Those who have such a conception will be certain, unquestioningly certain in their conception, since it is for them a given. For them the conception is simply of reality. It is completely on a par with seeing a painting as a painting. It is immediate. It is there, right in front of them, perceived as such. The structure once again fits the model. The view is not worked out by a process of reasoning. It is not that those with the conception that, say, women's bodies are for the delectation of men, are other than misleadingly said to believe this. If they are thought to believe this the belief will be thought to arise from the conception. The conception is prior to, and deeper than belief. There is no gap in the conception, between outer and inner. A woman's body is seen as for the delectation of men, that is how they are perceived. That is the immediate experience, as immediate as in the case of the painting.

Talk of evidence and observation is misplaced. Boys do not wear dresses, girls do. These are for many persons conceptual truths, not inductive inferences. Boys have short hair, girls long hair. These are more conceptual truths. The sight of a four-year-old girl with short hair but in a dress can, and does, induce stress for some people. The stress is conceptual. Short hair is classificatory, is part of the conception of being male. It is seen as male. The stress comes from conflicting conceptions, two conflicting "seeing-as" experiences.

But such deeply rooted conceptions, though they seem to succeed marvellously in providing certainty, do not justify this certainty. Such certainty in the cases I have canvassed is linked with established error. It fails to connect with reality. We cannot escape a gap, between how matters are conceived and how they are. Whether or not the outer is conceived of as expressive of the inner does not enable us to leap that gap. Such conceptualising provides psychological certainty, but that is compatible with profound, even pernicious error. That there were witches was questionable (and questioned) and false. That women are inferior is questionable, and questioned, and false. That blacks are inferior is questionable, and questioned, and false. That the other has a "soul" is questionable, not, perhaps other than pathologically, questioned, and true. But our certainty cannot properly come from the "attitudinal approach" to other minds. Our assurance needs to come from somewhere else. It is reassuring to have the analogical inference to other minds, waiting, faithfully, as always in the wings, as the various pretenders are banished from the stage.[5]

CHAPTER NINE

STRAWSON ON OTHER MINDS

The final attack that I shall consider on the analogical inference to other minds is that put forward by P.F. Strawson. The analogical inference to other minds is, of course, intended to be a solution to the problem of other minds. It is to provide a justification for believing in other minds. Strawson has an argument that this project is doomed from the start (or even before). His claim is that a solution or justification is impossible; 'the demand for it cannot be coherently stated'.[1] It is clear that Strawson's attack is not confined to the analogical inference to other minds. All arguments to other minds are supposed to fall prey to this onslaught.

Strawson's simple thought (as he puts it) is that I cannot ascribe experiences to myself unless I am also able to ascribe experiences to others. In order to do this I must be able to identify those others. I could not identify those others if private experiences were all I had to go on (see pp. 100 and 104).

A crucial claim is the first one, dubbed explicitly 'a very simple ... thought', and formulated initially as:

> ... it is a necessary condition of one's ascribing states of consciousness, experiences, to oneself, in the way one does, that one should also ascribe them, or be prepared to ascribe them, to others who are not oneself.[2]

This simple thought

> is a purely logical one: the idea of a predicate is correlative with that of a *range* of distinguishable individuals of which the predicate can be significantly, though not necessarily truly, affirmed.[3]

Strawson thinks the analogical inference falls foul of this principle. One cannot move from one's own experiences and their correlations to those of others since one is not entitled to speak of one's own experiences unless one already knows how to ascribe experiences to others:

> for unless he [the analogical arguer] already knows how to do this, he has no conception of *his own case*, or any *case*, i.e. any subject of experiences.[4]

Although neither a solution to the problem of other minds nor a justification for our belief in other minds is possible, given Strawson's 'simple thought', a consequent truth might be mistaken for just such a justification. When one ascribes experiences to others

> the behaviour-criteria one goes on are not just signs ... but are criteria of a logically adequate kind for the ascription ...[5]

This might be understood as indicating that such ascription is guaranteed meaningful but no more than that, which would leave many sceptics unmoved. But Strawson is not seeking to move sceptics about other minds. He is not offering a behaviour-criteria justification for belief in other minds. Strawson's insistence is that a justification is impossible, from which it follows that he is not offering one. There is no sceptical problem, since it cannot be stated, so there is no solution to that non-existent problem, consequently no justification for the relevant belief.

Strawson has two crucial claims, both needed to support his attack on arguments to other minds and, in particular, the analogical inference to other minds. The first is his principle that a predicate has to be ascribable to more that one individual. Why, though, should be accept this principle? Why do we need a *range* of *distinguishable* individuals? Could we not have an idea of ourselves as subjects of consciousness without knowing how to distinguish another subject of consciousness? More generally: why is it that we need more than one distinguishable individual? Why should not the notion of experiences as necessarily felt by a subject of consciousness not just occur to such a subject? If it is possible that it occur as the result of reasoning it seems an argument would be needed to support the claim that it could not just occur. Perhaps it could be programmed. Perhaps it could be arranged by a neurosurgeon. An argument is needed to counter these suggestions.

However, there would seem to be mere assertion, without argument. We are told we could not have an idea of ourselves unless we were able to distinguish others. But we lack supporting argument. Worse, as we shall see, Strawson's second crucial claim carries with it the notion that we (magically) just do acquire the capacity to self-ascribe and other ascribe on the basis of utterly distinct criteria in the two cases. If that is acceptable, why is it not acceptable just (magically?) to acquire the notion of ourselves without knowing how to distinguish others?

That second claim is that ascribability demands that there be 'ways of telling' whether an individual is a suitable candidate for the ascription of the predicate and these

> ways of telling must constitute logically adequate kinds of criteria for the ascription of the P-predicate.[6]

It seems that the logical adequacy is epistemic, given the flavour of his various remarks and, in a particular example, his remarks on depression and the fact that depression (in others) is observable. The adequacy, in any case, has to be epistemic, to be of any use in the immediate context of other minds (otherwise we should have at best established conditions for meaningful ascription, which would leave much epistemic work still to be done).

So understood, as epistemic, it seems straightforwardly implausible, confusing conceptual questions having to do with meaningful utterance with epistemological questions having to do with the rationality of beliefs expressed by utterances. To take just one example: the application of Strawson's principle to 'witches' would mean that the sceptical problem could not be raised. Given that 'witches' has a use in our language then there are to be logically adequate ways of telling whether an individual is a witch. If these ways are epistemic then meaningful claims that a person is a witch seem to carry a thoroughly dismaying epistemic guarantee. More rarified examples would be found, I believe, from time to time in science, where concepts are introduced and scientists then meditate on how they might test for the presence of entities postulated by the concepts. Strawson's principle would seem to guarantee not just that those scientists must succeed in their quest but that they would never be forced to embark on such a quest.[7]

The dubiousness of the general principle carries over into the particular case of other minds. The sceptical problem surely can be raised.

What justification is there for believing in minds? The complication that we have in the case of such a belief is the asymmetry between first person and third person ascriptions but the complication does not affect the immediate question. Why are the logically adequate criteria logically adequate? Why are they what they are? We surely need some explanation or we leave it wholly mysterious.

That seems to leave an opening for the analogical inference to other minds. So far there seems to be no barrier to bringing in the analogical inference to other minds after we have gratefully accepted what Strawson has to offer. If Strawson were to be interpreted as insisting merely on conditions for meaningful ascription then the analogical inference could be used to offer the required epistemic justification. Strawson's principle could be seen to license meaningful ascription. The analogical inference would license the belief. However, given that Strawson's principle has to be interpreted epistemically the *same possibility remains*. We can accept what Strawson has to offer and then use the analogical inference to other minds, either to dispel mystery, or to "improve" on the epistemic justification offered, or whatever.

The key notion is that the analogical arguer can accept Strawson and then say: but that leaves me puzzled, dissatisfied, or whatever; I want to use the analogical inference to other minds (perhaps they are nervous and like to have more arguments than they strictly need). There is nothing in Strawson's principle that prevents use of the analogical inference, *granted Strawson's principle*, and there is nothing in that principle that renders the analogical inference superfluous. If Strawson's principle gives us adequacy, perhaps we can do better.

What might Strawson's response be? The most directly relevant passage I know of is his discussion of what would follow if it were granted that the 'ways of telling' that a predicate could be ascribed never constituted 'logically adequate kinds of criteria'. Here is Strawson's comment:

> Then we should have to think of the relation between the ways of telling and what the P-predicate ascribes, or a part of what it ascribes, always in the following way: we should have to think of the ways of telling as *signs* of the presence, in the individual concerned, of this different thing, viz. the state of consciousness. But then we could only know that the way of telling was a sign of the presence of the different thing ascribed by the P-predicate, by the observation of correlations between the two. But this observation we could each make only in one case, viz. our own. And now we

> are back in the position of the defender of Cartesianism, who thought our way with it was too short. For what, now, does 'our own case' mean? There is no sense in the idea of ascribing states of consciousness to oneself, or at all, unless the ascriber already knows how to ascribe at least some states of consciousness to others. So he cannot argue in general 'from his own case' to conclusions about how to do this; for unless he already knows how to do this, he has no conception of *his own case*, or any *case*, i.e. any subject of experiences. Instead, he just has evidence that pain &c. may be expected when a certain body is affected in certain ways and not when others are. If he speculated to the contrary, his speculations would be immediately falsified.[8]

There is nothing in this that tells against my claim that a two-tiered approach to the problem is perfectly respectable. The analogical inferrer's conception of 'his own case' can be owed to Strawson's principle, and they can rejoice in whatever epistemic security that principle offers. Nothing prevents them then using the analogical inference to other minds or any other argument to other minds. Such arguments can only be attacked as superfluous and they can only be attacked as superfluous if the epistemic justification they afford is inferior to that afforded by Strawson.

So, Strawson's principle, even if it is true, does not make attempts to justify belief in other minds illegitimate. At best, arguments to other minds cannot proceed without Strawson, but they can proceed with him, and nothing he has claimed begins to suggest that they are not needed. If the principle's aim was to destroy, in particular, the analogical inference to other minds, then it has been, in a sense, self-defeating. Either it fails to persuade as a principle, or it can be used by those running that inference.

I have argued that Strawson's principle is dubious. I have then argued that even if it is granted arguments to other minds survive. The latter argument can be presented as a dilemma. Either the principle is a principle of meaningful ascription or it is to be put to epistemic use. If it is the former, then clearly it both leaves the epistemological issues untouched and can be accepted as a (helpful) preliminary to epistemological arguments. If it is the latter, then whether or not it works epistemologically remains an issue and other arguments might still be needed. Presumably if it works epistemologically it also works as a principle of meaningful ascription, so there seems no escaping that arguments to other minds can survive, if noway else, by embracing their attacker.

So far then, the principle is dubious, and even if it were successful, arguments to other minds would survive, by accepting the principle and then making their contribution. I now want to argue that the analogical inference to other minds survives in a stronger sense, in that it can meet Strawson's principle.

Plantinga has claimed that the condition on ascription (having a conception of the appropriate occasions for predicating experiences of others) can be met in this way: the analogical inferrer might think

> if N can predicate mental states of himself at time t, then, at t, N knows (or believes, thinks, etc.) that for any mental state M he predicates of himself, if he were to observe a correlation between a state B of his body and the presence of M, then any occasion on which he observes a body similar to his in state B is an appropriate occasion for predicating M of another subject of experience.[9]

But this suggestion would surely be rejected by Strawson. The problem is more than the fancifulness and the mysteriousness (where does this felicitous anticipation of reality come from?). What would be unacceptable to Strawson is the hypothetical nature of the conception. Strawson's principle demands that one be 'able to ascribe' experiences to others. At Plantinga's time t, the analogical inferrer simply cannot do so. They would be able to, if things turn out a certain way, but that is different from being able to do so.[10]

Bradley thinks the trick might be managed since I am entitled to speak of my body as opposed to another's.

> And someone might move from this to saying that *these* experiences (the experiences he knows about, or, as Strawson puts it, "all experience" - so far) I shall call 'mine', hypothesizing that other bodies have qualitatively similar ones uniquely related to them as these are to mine. That is, someone might succeed in introducing the distinction between his own experience and that of others by reference to the existing distinction between his own body and that of others. Hence he succeeds in side-stepping the objection that the signs-view is incompatible with the principle, by showing that the my experience-your experience distinction is possible within the framework of the signs-view.[11]

But surely Strawson would reject the opening claim: that I am entitled to speak of my body. All I can legitimately do is speak of this body and that body. Bradley conducts his discussion in terms of the following principle:

> if someone cannot say what it would be for an experience not to be his, then there is absolutely no point in saying that such an experience *is* his.[12]

What would be the point, then, of saying that a body was mine? If it was not, then it would presumably be another's. Unless I had a conception of myself as opposed to another there would be no content to this claim.

However, I think the trick *can* be managed and Bradley has pointed in the right direction. The most Strawson allows is that the solitary thinker could note that

> experiences, all experiences, stand in a special relation to body M, that body M is unique in just this way, that this is what makes body M unique among bodies.[13]

There is no entitlement to talk of my body. That is an additional claim. Such a thinker

> just has evidence that pain etc. may be expected when a certain body is affected in certain ways and not when others are. If he speculated to the contrary, his speculations would be immediately falsified.[14]

What if they are not so easily discouraged? They note that pain occurs in certain circumstances and is followed by certain behaviour on the part of body M. They decide that the circumstances in question cause the pain which in turn causes the behaviour. Those circumstances occur for body N followed by the behaviour. So they are presented with the circumstances and the behaviour without the pain. They have a puzzle in that the behaviour is explained for body M but not for body N. They find a similar puzzle in the case of itching and scratching.

The solution, though, is available. Pains and itches are such that there is pain and itching in body N's case as well as in M's but not this pain and this itching. What has to be the case if that is so? Pains and itches have to be such that they only exist when felt by a subject whose experiences they are. The associated body is that subject's. So that there is a subject S_1 for M and the solitary thinker is that subject and there is another subject for N. S_1 can ascribe pain then to S_1 and S_2, but differently. So S_1 and S_2 arrive together and the solitary thinker is one of them (and, in particular, themselves).

I have now argued that Strawson's basic principle is dubious, that arguments to other minds can survive that principle by accepting it as a preliminary to their own contribution, and that arguments from one's own case can accept the basic principle.

Strawson returned to his theme some years after its initial formulation.[15] Responding to, and following, Hide Ishiguro, he lists several elements in the essential make-up of persons, the most important being that:

> in order to be a person, one must see oneself and others *as* persons. We must *have* an operative concept of a person ... in order to *be* one.[16]

That one needs the concept would follow from the fact that you cannot see X as F unless you have the concept of an F. Strawson seems to regard seeing others as persons as equivalent to taking others to be persons. This is not so, and crucially not so epistemically. One may see X as F without even believing X to be F (never mind having a good reason to do so). If one takes X to be F one believes X to be F. Clearly Strawson believes others are persons. He is claiming that we must do so in order to be, ourselves, persons.

This later principle seems substantially the same as the earlier but purged of any epistemic import. What if we grant this principle? It is surely subject to the objections to the other principle (e.g. as before, I can still wonder whether it is rational to believe others are persons, and invoke some argument to other minds to satisfy this passing doubt). There are, in addition, new problems. Strawson has not established by his new principle that it is rational to believe others are persons, only, at best, that it is necessary in order to be a person. So, at best, it would be rational to believe others are persons in order to license one's belief that oneself is a person. But that is not the same as its being rational to believe others are persons.

Given that it is customary to think belief is not subject to the will, believing in others in order to believe in oneself seems odd. What if I wonder whether it is rational and, while so doing, put my belief "on ice"?

On Strawson's principle I have ceased to be a person. It would be more sensible to cease to believe in Strawson's principle.

Once again, and for the moment, finally, the analogical inference to other minds survives as a solution to the problem of other minds.

NOTES

NOTES TO INTRODUCTION

1 I owe these splendidly apposite quotations to John Wisdom who used them to prefix his contribution to the 'Symposium: Other Minds' (with J.L. Austin) *Proceedings of the Aristotelian Society*, XX (1946) (supplementary volume) pp. 122-87. I comment on these passages at the end of chapter one.

NOTES TO CHAPTER ONE

1 Tom Nagel, *The View from Nowhere* (OUP, 1986) p. 20.

2 *Ibid.*

3 Ludwig Wittgenstein, *Philosophical Investigations* (Blackwell, 1953) section 302.

4 *Op. cit.*, p. 20.

5 Colin McGinn, 'What is the Problem of Other Minds?', *Aristotelian Society. Proceedings, Supplementary Volume* No. 58 (1984) pp. 119-37. This is the second part of a symposium 'Consciousness and other minds'.

6 *Ibid.*, p. 136.

7 Christopher Peacocke, 'No Resting Place: A critical notice of *The View from Nowhere*', *Philosophical Review*, 98 (1989). See pp. 75-6.

8 Alvin Plantinga, *God and Other Minds* (Cornell, 1967) pp. 193-9.

9 It is worth pointing out that the quotation from Proust at the beginning of my Introduction talks of our lack of direct knowledge (of others) (the revised translation in Penguin Classics uses the same phrase). However the French phrase is 'connaissance directe'. I am sure from the surrounding text that the notion involved is direct observation (knowledge of) rather than the specific sense of direct knowledge invoked in my account (however tempting it is to think that one has been anticipated by Proust). Wisdom himself thought the 'peculiarity of the soul is not that it is visible to none but that it is visible only to one' (see my comments on him in chapter Four). The Proust passage is from *The Guermantes Way*, Part One (page 84 in the English

translation (Chatto and Windus, 1966) and p. 132 in the Flammarion edition of 1987).

NOTES TO CHAPTER TWO

1 D.M. Armstrong, *A Materialist Theory of the Mind* (Routledge, 1968) p. 125.

NOTES TO CHAPTER THREE

* Some of the material in this chapter appeared in an earlier version in my 'Other Minds and Theoretical Entities', *Australasian Journal of Philosophy*, vol.54 (1976) pp.158-61.

1 See Philip A. Ostien, 'God, Other Minds and the Inference to the Best Explanation', *Canadian Journal of Philosophy*, vol.4 (1974/1975) pp.149-62, Robert Pargetter, 'The Scientific Inference to Other Minds', *Australasian Journal of Philosophy*, vol.62 (1984) pp.158-63.

2 I would add to the more formal references (cited in footnote 1) the discussion by Douglas R. Hofstadter 'The Turing Test: A Coffee-house Conversation' and Daniel C. Dennett's 'Reflections' thereon (in their *The Mind's I* (Harvester Press, 1931) pp.69-95). This is stimulating and insightful, although informal. It is probably a useful example of the indirect way in which a commitment to SI often manifests itself.

3 Nathan Stemmer, 'The Hypothesis of Other Minds: Is it the Best Explanation?', *Philosophical Studies*, vol.51 (1987) pp.109-21, elaborates the considerations that favour the purely physical hypothesis. He calls himself an Interpretative Eliminativist but he might well be, or be prepared to be, part Reductionist, part Eliminativist.

4 So Central State Materialism and Functionalist Materialism can be added to Epiphenomenalism and Eliminativism to make the case against Dualists and Reductionist Materialists being able to use SI. A wide range of positions on the Mind-Body Problem deny that what causes human behaviour are mental states as traditionally understood.

5 It should be noted explicitly that the argument outlined to other minds is a hybrid of the traditional analogical inference to other minds and a scientific inference (to the best explanation).

NOTES TO CHAPTER FOUR

* This chapter has developed over the years out of A. Hyslop and F.C. Jackson, 'The Analogical Inference to Other Minds', *American Philosophical Quarterly*, vol.9 (1972) pp.168-76.

1 This point is elaborated later in this chapter.

2 We also believe, of course, that there is exactly one person to each living human body; but this is not something we should expect the analogical inference to other minds to establish.

3 R.I. Sikora, 'The Argument from Analogy is *not* an Argument for Other Minds', *American Philosophical Quarterly*, vol.14 (1977) pp.137-41.

4 'Wittgenstein's *Philosophical Investigations*' *in* V.C. Chappell (ed.), *The Philosophy of Mind* (Prentice-Hall, 1962).

5 See Donald Davidson, 'Psychology as philosophy', *in* J. Glover (ed.), *The Philosophy of Mind* (OUP, 1976), and also his 'Mental States', *in* L. Foster and J.W. Swanson (eds.), *Experience and Theory* (Duckworth, 1970); Paul Ziff, 'The Simplicity of Other Minds', *Journal of Philosophy*, vol.62 (1965) pp. 575-84; Sydney Shoemaker, 'Ziff's Other Minds', *Journal of Philosophy*, vol.62 (1965) pp. 587-9; Jaegwon Kim, 'On the Psycho-Physical Identity Theory', *American Philosophical Quarterly*, vol.3 (1966) pp. 227-35; David Lewis, 'An Argument for the Identity Theory', *Journal of Philosophy*, vol.63 (1966), pp. 17-25.

6 My response to Sikora is a revised version of my 'A Multiple Case Inference and Other Minds', *Australasian Journal of Philosophy*, vol.57 (1979) pp. 330-6.

7 Don Locke, *Myself and Others* (OUP, 1968) p. 49.

8 Norman Malcolm, 'Knowledge of Other Minds', *in* V.C. Chappell (ed.), *The Philosophy of Mind* (Prentice-Hall, 1962) pp. 151-9 (see p. 152).

9 J. Day, *Inductive Probability* (Routledge, 1962) p. 64.

10 R.J. Pargetter, 'The Scientific Inference to Other Minds', *Australasian Journal of Philosophy*, vol.62 (1984) p. 160.

11 *Australasian Journal of Philosophy*, vol.54 (1976) pp. 158-61.

12 *Op. cit.*

13 'In Defense of Type Materialism', *Synthese*, vol.59 (1984) pp. 295-320. See pp. 310-14.

14 Carl Wellman, 'Our Criteria for Third Person Psychological Sentences', *Journal of Philosophy*, vol.58 (1961) pp. 281-93 (see pp. 292-3).

15 Don Locke, *op.cit.*, pp. 49-50.

16 Don Locke, 'Just What is Wrong with the Argument from Analogy?', *Australasian Journal of Philosophy*, vol.51 (1973) pp. 153-6.

17 My response to Don Locke's response is essentially that to be found in Alec Hyslop and Frank Jackson, 'A Reply to Don Locke', *Australasian Journal of Philosophy*, vol.53 (1975) pp. 68-9.

18 The *locus classicus* is Gilbert Ryle, *The Concept of Mind* (Hutchinson, 1949) see p. 52.

19 *Ibid.*

20 Stuart Hampshire, 'The Analogy of Feeling', *Mind*, vol.61 (1952) pp. 1-12 (see p. 4).

21 Norman Malcolm, *op.cit.*, p. 153.

22 V.C. Chappell (ed.), *The Philosophy of Mind* (Prentice-Hall, 1962) p. 3.

23 Peter Alexander, 'Other People's Experiences', *Aristotelian Society Proceedings*, vol. 51 (1950/51) pp. 24-46 (see pp. 43-4). Alexander is considering, without definitely endorsing, this objection.

24 Alvin Plantinga, *God and Other Minds* (Cornell, 1967) p. 248.

25 John Wisdom, *Other Minds*, 2nd ed. (Blackwell, 1965).

26 John Wisdom, 'The Concept of Mind', *Other Minds, op. cit.*, pp. 232-44 (see p. 237).

27 Plantinga, *op. cit.*, pp. 245-68. See also his 'Induction and Other Minds', *Review of Metaphysics*, vol.19 (1965-1966) pp. 441-61 and Michael Slote's reply in *Review of Metaphysics*, vol.20 (1966-1967) pp. 341-60 and Plantinga's reply to Slote in the same issue. I discuss Plantinga's criticism of the analogical inference to other minds because, apart from being an unusually penetrating one, it seems to me that Slote's reply to it fails.

28 I owe this suggestion to John Fox.

29 See Hector Neri-Castañeda, 'Consciousness and Behaviour', *in Intentionality, Minds and Perception,* ed., Castañeda (Wayne State, 1966) pp. 132-3.

30 Karl Ameriks, 'Plantinga and Other Minds', *Southern Journal of Philosophy*, vol.16 (1973) pp. 285-91.

31 *Ibid.*, p. 286.

32 It should be noted that Ameriks is defending the analogical inference to other minds in this article by constructing a version which complies with his principles.

33 Sydney Shoemaker, *Self-knowledge and self-identity* (Cornell, 1963) p. 168.

NOTES TO CHAPTER FIVE

* This chapter is a revised, and enlarged, version of my 'Criteria and Other Minds', *Australasian Journal of Philosophy*, vol.51 (1973) pp. 105-14.

1 Norman Malcolm, *Dreaming* (Routledge, 1959); together with his 'Knowledge of Other Minds', *Journal of Philosophy*, vol.55 (1958) pp. 969-78; and his 'Wittgenstein's *Philosophical Investigations*', *Philosophical Review*, vol.63 (1954) pp. 530-59.

2 Bruce Aune, 'Feelings, Moods and Introspection', *Mind*, vol.62 (1963) pp. 187-208. Sydney Shoemaker, *Self-knowledge and Self-identity* (Cornell, 1963).

3 Norman Malcolm, 'Wittgenstein's *Philosophical Investigations*', *op.cit.*, p. 544.

4 *Ibid.*, pp. 546-7.

5 'Knowledge of Other Minds', *op.cit.*, p. 973.

6 'Wittgenstein's *Philosophical Investigations*', *op.cit.*, p. 544.

7 *Dreaming*, *op.cit.*, p. 35.

8 John T. Saunders and Donald F. Henze, *The Private Language Problem* (Random House, 1967) p. 111.

9 W.D. Glasgow and G.W. Pilkington, 'Other Minds on Evidential Necessity', *Mind*, vol.79 (1970) p. 435.

10 Robert C. Coburn, 'Persons and Psychological Concepts', *American Philosophical Quarterly*, vol.4 (1967) p. 211.

11 Sydney Shoemaker, *op. cit.*, pp. 3-4.

12 Bruce Aune, *op. cit.*, p. 203.

13 Robert C. Coburn, *op. cit.*, p. 211.

14 Sydney Shoemaker, *op. cit.*, pp. 186-93.

15 Alvin Plantinga, *God and Other Minds* (Cornell, 1967) pp. 226-7.

16 Bernard Williams, 'Knowledge and Meaning in the Philosophy of Mind', *Philosophical Review*, vol.77 (1968) pp. 225-8.

17 Sydney Shoemaker, *op. cit.*, pp. 192-3.

18 W. Gregory Lycan, 'Non-inductive Evidence: Recent Work on Wittgenstein's "Criteria"', *American Philosophical Quarterly*, vol.8 (1971) pp. 114-5.

19 *Op.cit.*

20 Sydney Shoemaker, 'Critical Study of Don Locke *Myself and Others*', *Philosophical Quarterly*, vol.19 (1969) p. 273.

21 Gilbert Ryle, *The Concept of Mind* (Hutchinson, 1949) p. 109.

22 Bruce Aune, *op. cit.*, p. 204.

23 Bruce Aune, *Knowledge, Mind and Nature* (Random House, 1967) p. 279.

24 I understand that the most that can be claimed here is that sometimes, when burned, one gets a burning pain, and that is the basis for the characterisation.

25 Bruce Aune, *op. cit.*, p. 202.

26 Bruce Aune, *Knowledge, Mind, and Nature*, p. 131.

27 John McDowell, 'Criteria, Defeasibility, and Knowledge', *Proceedings of the British Academy*, LXVIII (1982) pp. 455-79.

28 Paul Robinson, 'McDowell against Criterial Knowledge', *Ratio*, NS. 4 (1991) pp. 59-75.

29 *Ibid.*, p. 60.

30 *Ibid.*, p. 74.

NOTES TO CHAPTER SIX

1 Robinson Krusoe is the supposed speaker of the logically private "language".

2 See the comments on Malcolm in Chapter Five.

3 Ludwig Wittgenstein, *Philosophical Investigations* (Oxford, Blackwell, 1953) Paragraph 258.

4 Norman Malcolm, 'Wittgenstein's *Philosophical Investigations*' *in* V.C. Chappell (ed.), *The Philosophy of Mind* (Prentice-Hall, Englewood Cliffs, 1962) pp. 74-100. See p. 76.

5 *Op. cit.*, Paragraph 293.

6 A.J. Ayer, 'The concept of a person' *in The Concept of a Person* (Macmillan, 1963). See pp. 106-8.

NOTES TO CHAPTER SEVEN

* Kripstein is Wittgenstein as interpreted by Kripke.

1 Saul A. Kripke, *Wittgenstein on Rules and Private Language* (Blackwell, 1982) p. 60.

2 Kripke, *op. cit.*, p. 23.

3 *Ibid.*, p. 37.

4 *Ibid.*, p. 43.

5 *Ibid.*, p. 51.

6 *Ibid.*, p. 60. I take it that such a language does not fail directly in that sense might conceivably be made of rule-following by reference to other than other utterers.

7 Kripstein draws an analogy with Hume's treatment of causation.

8 *Op. cit.*, pp. 94-5.

9 *Ibid.*, p. 112.

10 Crispin Wright, *Wittgenstein on the Foundations of Mathematics* (Duckworth, 1980) p. 222. Wright seems to have broadly the same view of Wittgenstein on rule-following that Kripke is putting forward.

11 *Ibid.*, pp. 103-4.

12 Kripke, *op. cit.*, p. 110.

13 Wright, *op. cit.*, p. 106.

14 Kripke, *op. cit.*, pp. 102-4.

15 Wright, *op. cit.*, p. 218.

16 *Ibid.*, p. 219.

17 *Ibid.*

18 Stanley Cavell, *Must we Mean What we Say*? (CUP, 1976) p. 52.

19 J.R. Cameron, Review of Crispin Wright, *Foundations of Mathematics, op. cit., Philosophical Books*, vol.23 (1982) pp. 86-90.

20 *Ibid.*, p. 89.

21 *Ibid.*

22 *Ibid.*

23 *Ibid.*

24 *Ibid.*, p. 90.

25 Steven H. Holtzman and Christopher M. Leich (eds.), *Wittgenstein: to Follow a Rule* (Routledge, 1981) Introductory essay.

26 *Ibid.*, p. 4.

27 Kripke, *op. cit.*, p. 110.

28 *Op. cit.*, p. 4.

29 *Ibid.*, p. 5.

30 *Ibid.*, p. 13.

31 I take this response to be common to Simon Blackburn, 'The Individual Strikes Back', *Synthese*, vol.58 (1984) pp. 281-301; Warren Goldfarb, 'Kripke on Wittgenstein on Rules', *Journal of Philosophy*, vol.82 (1985) pp. 471-88; Paul Hoffman, 'Kripke on Private Language', *Philosophical Studies*, vol.47 (1985) pp. 23-8; Harry A. Lewis, 'Content and Community', *The Aristotelian Society*, supp. vol.59 (1985) pp. 177-96; Colin McGinn, *Wittgenstein on Meaning* (Blackwell, 1984) p. 168.

32 *In* Holtzman and Leich, *op. cit.*, p. 183.

33 *Op. cit.*, p. 52.

34 Kripke, *op. cit.*, p. 104 (fn. 83).

35 *Ibid.*, p. 103 (fn. 83).

36 Kripke, *op. cit.*, p. 51.

37 *Ibid.*, p. 51.

38 *Ibid.*, p. 53.

39 *Ibid.*, p. 52.

40 *Ibid.*, p. 55.

NOTES TO CHAPTER EIGHT

1 M.R.M. Ter Hark, 'The Development of Wittgenstein's Views about the Other Minds Problem', *Synthese* vol.87 (1991) pp. 227-53.

2 It should be noted that mutual behaviourism (holding of ourselves and others) (a) seems never to have been held by Wittgenstein and (b), is in any case, subject to the particular asymmetry I have been insisting on, as argued in chapter two.

3 See Ter Hark, *op. cit.*, section 3 'Pretending' and M.B. Hintikka and J. Hintikka, *Investigating Wittgenstein* (Blackwell, 1986) pp. 279-84.

4 John McDowell, 'On "The Reality of the Past"', in Christopher Hookway and Philip Pettit (eds.), *Action and Interpretation* (Cambridge University Press, 1978) pp. 127-44. In his 'Criteria, Defeasibility, and Knowledge' (*op. cit.*) he does not advocate direct awareness of another's inner states. However, there is a tantalising discussion on pp. 472-3 (and footnote 1 on p. 473) which might be understood to contain a hint of the "attitudinal approach" as I have elaborated it. It should be noted, once again, that direct observation of the inner selves of others would not avoid the problem of other minds. What is needed is direct knowledge *that* other human figures have inner selves.

5 It is likely that some have "attached" the "attitudinal approach" to whatever "dissolution" of the other minds problem they think successful. The "dissolution" demonstrates the problem's incoherence. The "attitudinal approach" describes our actual views of others. The work is done by the "dissolution". Such a strategy would seem to be less attractive than claiming that it is rational to believe in others because it is incoherent to do otherwise.

NOTES TO CHAPTER NINE

1 P.F. Strawson, *Individuals* (London: Methuen, 1959) p. 112.
2 *Ibid.*, p. 99.
3 *Ibid.*
4 *Ibid.*, p. 106.
5 *Ibid.*
6 *Ibid.*, p. 105. A P-predicate is a predicate that applies to persons and ascribes states of consciousness.
7 I understand that neutrinos and quarks are such entities. They were postulated to explain certain (curious) observations. Scientists took some time to work out how they might "detect" these entities.
8 *Op. cit.*, pp. 105-6.
9 Alvin Plantinga, *God and Other Minds* (Cornell University Press, 1967) pp. 210-11.
10 The matter is complicated by Strawson's footnote on p. 99, which is where, I take it, Plantinga found his condition on ascription (in terms of having a conception of 'the appropriate occasions ...'). Unravelling all that would be time-confusing but would not, I think, protect Plantinga. Briefly, having a conception of the appropriate occasions for ascribing is distinct from having a conception of how things would have to be if one were to be able to ascribe.
11 M.C. Bradley, 'Mr. Strawson and scepticism', *Analysis* (1959) pp. 14-19 (p. 15).
12 *Ibid.*
13 *Op. cit.*, p. 101.
14 *Ibid.*, p. 106.
15 *In* Zak van Straaten (ed.), *Philosophical Subjects: Essays presented to P.F. Strawson* (Clarendon Press, 1980). Strawson's reply to Hide Ishiguro's article is the occasion.

16 *Ibid*., p. 272.

REFERENCES

ALEXANDER, P. 'Other People's Experiences', *Aristotelian Society Proceedings*, vol. 51 (1950/1951) pp. 24-46.

AMERIKS, K. 'Plantinga and Other Minds', *Southern Journal of Philosophy*, vol. 16 (1973) pp. 285-91.

ARMSTRONG, D.M. *A Materialist Theory of the Mind* (Routledge, 1968).

AUNE, B. 'Feelings, Moods and Introspection', *Mind*, vol. 62 (1963) pp. 187-208.

___________. *Knowledge, Mind and Nature* (Random House, 1967).

AYER, A.J. 'The Concept of a Person', *in* A.J. Ayer, *The Concept of a Person* (Macmillan, 1963).

BLACKBURN, S. 'The Individual Strikes Back', *Synthese*, vol. 58 (1984) pp. 281-301.

BRADLEY, M.C. 'Mr. Strawson and Scepticism', *Analysis*, vol. 20 (1959/1960) pp. 14-19.

CAMERON, J.R. Review of Crispin Wright, *Wittgenstein on the Foundations of Mathematics* (Duckworth, 1982).

CASTAÑEDA, H-N. 'Consciousness and Behaviour', in H-N. Castañeda (ed.), *Intentionality, Minds and Perception* (Wayne State, 1966).

CAVELL, S. *Must We Mean What We Say?* (Cambridge University Press, 1976).

CHAPPELL, V.C. (ed.) *The Philosophy of Mind* (Prentice-Hall, 1962).

COBURN, R.C. 'Persons and Psychological Concepts', *American Philosophical Quarterly*, vol. 4 (1967) pp. 208-21.

DAVIDSON, D. 'Mental States', *in* L. Foster and J.W. Swanson (eds.), *Experience and Theory* (Duckworth, 1970).

___________. 'Psychology as Philosophy', in J. Glover (ed.), *The Philosophy of Mind* (Oxford University Press, 1976).

DAY, J.P. *Inductive Probability* (Routledge, 1961).

GLASGOW, W.D. and PILKINGTON, G.W. 'Other Minds on Evidential Necessity', *Mind*, vol. 79 (1970) pp. 431-5.

GOLDFARB, W. 'Kripke on Wittgenstein on Rules', *Journal of Philosophy*, vol. 82 (1985) pp. 471-88.

HAMPSHIRE, S. 'The Analogy of Feeling', *Mind*, vol. 61 (1952) pp. 1-12.

HINTIKKA, M.B. and HINTIKKA, J. *Investigating Wittgenstein* (Blackwell, 1986).

HOFFMAN, P. 'Kripke on Private Language', *Philosophical Studies*, vol. 47 (1985) pp. 23-28.

HOLTZMAN, S.H. and LEICH, C.M. (eds.) *Wittgenstein: to Follow a Rule* (Routledge, 1981).

HYSLOP, A. 'Criteria and Other Minds', *Australasian Journal of Philosophy*, vol. 51 (1973) pp. 105-114.

___________. 'Other Minds and Theoretical Entities', *Australasian Journal of Philosophy*, vol. 54 (1976) pp. 158-161.

___________. 'A Multiple Case Inference and Other Minds', *australasian Journal of Philosophy*, vol. 57 (1979) pp. 330-336.

HYSLOP, A. and JACKSON, F.C. 'The Analogical Inference to Other Minds', *American Philosophical Quarterly*, vol. 9 (1972) pp. 168-176.

___________. 'A Reply to Don Locke', *Australasian Journal of Philosophy*, vol. 53 (1975) pp. 68-69.

KIM, J. 'On the Psycho-physical Identity Theory', *American Philosophical Quarterly*, vol. 3 (1966) pp. 227-35.

KRIPKE, S.A. *Wittgenstein on Rules and Private Language* (Blackwell, 1982).

LEWIS, D. 'An Argument for the Identity Theory', *Journal of Philosophy*, vol. 63 (1966) pp. 17-25.

LEWIS, H.A. 'Content and Community', *The Aristotelian Society*, Supp. vol. 59 (1985) pp. 177-96.

LOCKE, D. *Myself and Others* (Oxford University Press, 1968).

___________. 'Just What is Wrong with the Argument from Analogy?', *Australasian Journal of Philosophy*, vol. 51 (1973) pp. 153-6.

LYCAN, W.G. 'Non-inductive Evidence: Recent Work on Wittgenstein's "Criteria"', *American Philosophical Quarterly*, vol. 8 (1971) pp. 109-25.

MALCOLM, N. *Dreaming* (Routledge, 1959).

___________. 'Knowledge of Other Minds', *in* V.C. Chappell (ed.), *The Philosophy of Mind* (Prentice-Hall, 1962).

___________. 'Wittgenstein's *Philosophical Investigations*', *in* V.C. Chappell (ed.), *The Philosophy of Mind* (Prentice-Hall, 1962).

McDOWELL, J. 'Criteria, Defeasibility, and Knowledge', *British Academy Proceedings*, vol. 68 (1982) pp. 455-79.

___________. 'On "The Reality of the Past"', *in* Christopher Hookway and Philip Pettit (eds.), *Action and Interpretation* (Cambridge University Press, 1978) pp. 127-44.

McGINN, C. 'What is the Problem of Other Minds?', *Aristotelian Society Proceedings*, Supp. vol. 58 (1984) pp. 119-37.

____________. *Wittgenstein on Meaning* (Blackwell, 1984).

NAGEL, T. *The View From Nowhere* (Oxford University Press, 1986).

OSTIEN, P.A. 'God, Other Minds and the Inference to the Best Explanation', *Canadian Journal of Philosophy*, vol. 4 (1974/1975) pp. 149-62.

PARGETTER, R. 'The Scientific Inference to Other Minds', *Australasian Journal of Philosophy*, vol. 62 (1984) pp. 158-63.

PEACOCKE, C. No Resting Place: a Critical Notice of *The View From Nowhere* [Tom Nagel], *Philosophical Review*, vol. 98 (1989) pp. 65-82.

PLANTINGA, A. 'Induction and Other Minds', *Review of Metaphysics*, vol. 19 (1965/1966) pp. 441-61.

____________. *God and Other Minds* (Cornell, 1967).

PROUST, M. *Le côté de Guermantes* (Flammarion, 1987) and (an English translation) *The Guermantes Way* (Chatto and Windus, 1966).

ROBINSON, P. 'McDowell Against Criterial Knowledge', *Ratio* (New series) vol. 4 (1991) pp. 59-75.

RYLE, G. *The Concept of Mind* (Hutchinson, 1949).

SAUNDERS, J.T. and HENZE, D.F. *The Private Language Problem* (Random House, 1967).

SHOEMAKER, S. 'Critical Study of Don Locke *Myself and Others*', *Philosophical Quarterly*, vol. 19 (1969) pp. 272-9.

____________. *Self-knowledge and Self-identity* (Cornell, 1963).

____________. 'Ziff's Other Minds', *Journal of Philosophy*, vol. 62 (1965) pp. 587-9.

SIKORA, R.I. 'The Argument from Analogy is *not* an Argument for Other Minds', *American Philosophical Quarterly*, vol. 14 (1977) pp. 137-41.

STEMMER, N. 'The Hypothesis of Other Minds: is it the Best Explanation?', *Philosophical Studies*, vol. 51 (1987) pp. 109-21.

STRAATEN, Z. von (ed.) *Philosophical Subjects: Essays Presented to P.F. Strawson* (Clarendon Press, 1980).

STRAWSON, P.F. *Individuals* (Methuen, 1959).

TER HARK, M.R.M. 'The Development of Wittgenstein's Views about the Other Minds Problem', *Synthese*, vol. 87 (1991) pp. 227-53.

WELLMAN, C. 'Our Criteria for Third-Person Psychological Sentences', *Journal of Philosophy*, vol. 58 (1961) pp. 281-93.

WILLIAMS, B. 'Knowledge and Meaning in the Philosophy of Mind', *Philosophical Review*, vol. 77 (1968) pp. 225-8.

WISDOM, J. *Other Minds*, 2nd edition (Blackwell, 1965).

WITTGENSTEIN, L. *Philosophical Investigations* (Blackwell, 1953).

WRIGHT, C. *Wittgenstein on the Foundations of Mathematics* (Duckworth, 1980).

ZIFF, P. 'The Simplicity of Other Minds', *Journal of Philosophy*, vol. 62 (1965) pp. 575-84.

INDEX

agreement
see community agreement
Alexander, P., 63
Ameriks, K., 68-69
analogical inference
see inference, analogical
Armstrong, D., 79
asymmetrical knowledge, 6-7, 17, 19-22, 30, 34-35, 71, 96, 121-126, 134
attitudinal approach to other minds, 121-130
Aune, B., 79, 81-83, 92-93, 145
Ayer, 42, 102

beetle in the box
see Wittgenstein, beetle in the box
behaviourism, 16, 18-19, 21-22, 25, 88, 101-102, 121-125, 128, 148
Blackburn, S., 116, 148
Bradley, M.C., 136-137

Cameron, J., 112-113
Campbell, K., 23
Canute, 88
Castaneda, H., 144
Cavell, S., 111-112, 116-117
central-state materialism
see materialism, central-state
Chappell, V., 62
Coburn, R., 77, 145
community agreement, 108-120
conventionality and criteria
see criteria, conventionality
criteria, 3, 71-91, 109, 126-128, 132-134
conventionality, 87-91
defeasibility, 84-91
inner process/outer criterion, 72-73

Dalmation language, 93
Davidson, D., 143
Day, J., 143

defeasibility
see criteria, defeasibility
definition, ostensive
see ostensive definition
Dennett, D., 142
Descartes, 115
direct knowledge
see knowledge, direct
dualism, 15, 18, 22-23, 25-26, 29-38
epiphenomenalist, 31-32, 36-37
interactionist, 32

E-type sensation
see private language argument (old)
eliminative materialism
see materialism, eliminative
epiphenomenalism
see dualism, epiphenomenalist

family resemblance, 107
form of life, 109
Fox, J., 144
functionalism, 96-98, 101-102
see also materialism, functionalist
functionalist materialism
see materialism, functionalist

Glasgow, W., 145
Goldfarb, W., 148

Hampshire, S., 61-62
Henze, D., 145
Hill, C., 54
Hintikka, J., 148
Hintikka, M., 148
Hoffman, P., 148
Hofstadter, D., 142
Holtzman, S., 113-115
Hume, 147
hypothetic inference
see inference, scientific
Hyslop, A., 54, 144

idealism, 15, 18
imagination, 10-13
incorrigibility thesis, 14
indirect knowledge
    see asymmetrical knowledge
inference
    analogical, 2-3, 7, 31, 36-38, 41-70, 81, 94-95, 121-125, 128, 130-132, 134-136, 139, 142
    one case, 3, 19-22, 25-27, 31-39, 41-54
    reliable, 62-63
    scientific, 2-3, 29-43, 72, 95
    uncheckable, 41, 54-70
inner process
    see criteria, inner process/outer criterion
interactionism
    see dualism, interactionist
Ishiguro, H., 138

Jackson, F.C., 25, 54, 144

Kim, J., 143
knowledge, asymmetrical
    see asymmetrical knowledge
knowledge, direct/indirect
    see asymmetrical knowledge
Kripke, S., 105-120

language, private
    see private language
Leich, C., 113-115
Lewis, D., 143
Lewis, H., 148
Locke, D., 51, 58-60
Lycan, W., 78

Malcolm, N., 13, 47, 51-52, 61-62, 73-77, 84, 97, 110, 146
materialism, 15-16, 24, 31-32, 102
    central-state, 17-18, 32, 34
    eliminative, 15, 18, 20-22, 24, 29, 33-34, 42
    functionalist, 15, 17-18, 21-26, 30-32, 34-35, 38, 96-97, 102
    reductionist, 31-35, 38, 102
McDowell, J., 84-91, 128
McGinn, C., 11-13, 148

Nagel, T., 8-10

observability, 5-6, 16
ostensive definition 108
Ostien, P., 142

Pargetter, R., 142-143
Peacocke, C., 13
Pilkington, G., 145
Plantinga, A., 13, 63, 66-68, 78, 136
predicates, 1310139
pretence, 127-129
private language, 3, 24, 93-120
private language argument (new) 105-120
private language argument (old) 93-103
    E-type sensation, 98-101
Proust, 1, 141-142
psychophysical correlations, 77-79

qualia, 18-22

reductionist materialism
    see materialism, reductionist
resemblance, family
    see family resemblance
Robinson Crusoe, 93-95, 102, 109, 113
Robinson, P., 88-90
rule-following, 99-100, 105-120
Ryle, G., 81, 144

Saunders, J., 145
scientific inference
    see inference, scientific
self-intimation thesis, 12
Shoemaker, S., 69-70, 78-79, 84, 89, 143, 145
Sikora, R., 43-50
Slote, M., 144
Stemmer, N., 142
Straaten, Z., 150
Strawson, P., 131-139

Ter Hark, M., 121, 148

theoretical entities
see inference, scientific
Thatcher, M., 129
theories of mind, 19-25, 34-36, 39, 96
transcendental argument, 23-25

Udina, A., 93, 102
uncheckability
see inference, uncheckable
uniformity of nature, 53
unobservable entities
see inference, scientific

vagueness, 107
verificationism, 30
verificationist theory of meaning, 25

Wellman, C., 57
Williams, B., 78
Wisdom, J., 64-66, 141
Wittgenstein, 8-9, 89-101, 110, 114, 121
beetle in the box, 100-102
Wright, C., 110-111, 147

Ziff, P., 143

65. H. E. Kyburg, Jr., *The Logical Foundations of Statistical Inference.* 1974
ISBN 90-277-0330-2; Pb 90-277-0430-9
66. M. Grene, *The Understanding of Nature.* Essays in the Philosophy of Biology. [Boston Studies in the Philosophy of Science, Vol. XXIII] 1974
ISBN 90-277-0462-7; Pb 90-277-0463-5
67. J. M. Broekman, *Structuralism: Moscow, Prague, Paris.* Translated from German. 1974 ISBN 90-277-0478-3
68. N. Geschwind, *Selected Papers on Language and the Brain.* [Boston Studies in the Philosophy of Science, Vol. XVI] 1974 ISBN 90-277-0262-4; Pb 90-277-0263-2
69. R. Fraissé, *Course of Mathematical Logic* – Volume 2: *Model Theory.* Translated from French. 1974 ISBN 90-277-0269-1; Pb 90-277-0510-0
(For *Volume 1* see under No. 54)
70. A. Grzegorczyk, *An Outline of Mathematical Logic.* Fundamental Results and Notions explained with all Details. Translated from Polish. 1974
ISBN 90-277-0359-0; Pb 90-277-0447-3
71. F. von Kutschera, *Philosophy of Language.* 1975 ISBN 90-277-0591-7
72. J. Manninen and R. Tuomela (eds.), *Essays on Explanation and Understanding.* Studies in the Foundations of Humanities and Social Sciences. 1976
ISBN 90-277-0592-5
73. J. Hintikka (ed.), *Rudolf Carnap, Logical Empiricist.* Materials and Perspectives. 1975 ISBN 90-277-0583-6
74. M. Čapek (ed.), *The Concepts of Space and Time.* Their Structure and Their Development. [Boston Studies in the Philosophy of Science, Vol. XXII] 1976
ISBN 90-277-0355-8; Pb 90-277-0375-2
75. J. Hintikka and U. Remes, *The Method of Analysis.* Its Geometrical Origin and Its General Significance. [Boston Studies in the Philosophy of Science, Vol. XXV] 1974 ISBN 90-277-0532-1; Pb 90-277-0543-7
76. J. E. Murdoch and E. D. Sylla (eds.), *The Cultural Context of Medieval Learning.* [Boston Studies in the Philosophy of Science, Vol. XXVI] 1975
ISBN 90-277-0560-7; Pb 90-277-0587-9
77. S. Amsterdamski, *Between Experience and Metaphysics.* Philosophical Problems of the Evolution of Science. [Boston Studies in the Philosophy of Science, Vol. XXXV] 1975 ISBN 90-277-0568-2; Pb 90-277-0580-1
78. P. Suppes (ed.), *Logic and Probability in Quantum Mechanics.* 1976
ISBN 90-277-0570-4; Pb 90-277-1200-X
79. H. von Helmholtz: *Epistemological Writings. The Paul Hertz / Moritz Schlick Centenary Edition of 1921 with Notes and Commentary by the Editors.* Newly translated from German by M. F. Lowe. Edited, with an Introduction and Bibliography, by R. S. Cohen and Y. Elkana. [Boston Studies in the Philosophy of Science, Vol. XXXVII] 1975 ISBN 90-277-0290-X; Pb 90-277-0582-8
80. J. Agassi, *Science in Flux.* [Boston Studies in the Philosophy of Science, Vol. XXVIII] 1975 ISBN 90-277-0584-4; Pb 90-277-0612-2
81. S. G. Harding (ed.), *Can Theories Be Refuted?* Essays on the Duhem-Quine Thesis. 1976 ISBN 90-277-0629-8; Pb 90-277-0630-1
82. S. Nowak, *Methodology of Sociological Research.* General Problems. 1977
ISBN 90-277-0486-4
83. J. Piaget, J.-B. Grize, A. Szemińska and V. Bang, *Epistemology and Psychology of Functions.* Translated from French. 1977 ISBN 90-277-0804-5

84. M. Grene and E. Mendelsohn (eds.), *Topics in the Philosophy of Biology.* [Boston Studies in the Philosophy of Science, Vol. XXVII] 1976
ISBN 90-277-0595-X; Pb 90-277-0596-8
85. E. Fischbein, *The Intuitive Sources of Probabilistic Thinking in Children.* 1975
ISBN 90-277-0626-3; Pb 90-277-1190-9
86. E. W. Adams, *The Logic of Conditionals.* An Application of Probability to Deductive Logic. 1975 ISBN 90-277-0631-X
87. M. Przełęcki and R. Wójcicki (eds.), *Twenty-Five Years of Logical Methodology in Poland.* Translated from Polish. 1976 ISBN 90-277-0601-8
88. J. Topolski, *The Methodology of History.* Translated from Polish by O. Wojtasiewicz. 1976 ISBN 90-277-0550-X
89. A. Kasher (ed.), *Language in Focus: Foundations, Methods and Systems.* Essays dedicated to Yehoshua Bar-Hillel. [Boston Studies in the Philosophy of Science, Vol. XLIII] 1976 ISBN 90-277-0644-1; Pb 90-277-0645-X
90. J. Hintikka, *The Intentions of Intentionality and Other New Models for Modalities.* 1975 ISBN 90-277-0633-6; Pb 90-277-0634-4
91. W. Stegmüller, *Collected Papers on Epistemology, Philosophy of Science and History of Philosophy.* 2 Volumes. 1977 Set ISBN 90-277-0767-7
92. D. M. Gabbay, *Investigations in Modal and Tense Logics with Applications to Problems in Philosophy and Linguistics.* 1976 ISBN 90-277-0656-5
93. R. J. Bogdan, *Local Induction.* 1976 ISBN 90-277-0649-2
94. S. Nowak, *Understanding and Prediction.* Essays in the Methodology of Social and Behavioral Theories. 1976 ISBN 90-277-0558-5; Pb 90-277-1199-2
95. P. Mittelstaedt, *Philosophical Problems of Modern Physics.* [Boston Studies in the Philosophy of Science, Vol. XVIII] 1976 ISBN 90-277-0285-3; Pb 90-277-0506-2
96. G. Holton and W. A. Blanpied (eds.), *Science and Its Public: The Changing Relationship.* [Boston Studies in the Philosophy of Science, Vol. XXXIII] 1976
ISBN 90-277-0657-3; Pb 90-277-0658-1
97. M. Brand and D. Walton (eds.), *Action Theory.* 1976 ISBN 90-277-0671-9
98. P. Gochet, *Outline of a Nominalist Theory of Propositions.* An Essay in the Theory of Meaning and in the Philosophy of Logic. 1980 ISBN 90-277-1031-7
99. R. S. Cohen, P. K. Feyerabend, and M. W. Wartofsky (eds.), *Essays in Memory of Imre Lakatos.* [Boston Studies in the Philosophy of Science, Vol. XXXIX] 1976
ISBN 90-277-0654-9; Pb 90-277-0655-7
100. R. S. Cohen and J. J. Stachel (eds.), *Selected Papers of Léon Rosenfield.* [Boston Studies in the Philosophy of Science, Vol. XXI] 1979
ISBN 90-277-0651-4; Pb 90-277-0652-2
101. R. S. Cohen, C. A. Hooker, A. C. Michalos and J. W. van Evra (eds.), *PSA 1974. Proceedings of the 1974 Biennial Meeting of the Philosophy of Science Association.* [Boston Studies in the Philosophy of Science, Vol. XXXII] 1976
ISBN 90-277-0647-6; Pb 90-277-0648-4
102. Y. Fried and J. Agassi, *Paranoia.* A Study in Diagnosis. [Boston Studies in the Philosophy of Science, Vol. L] 1976 ISBN 90-277-0704-9; Pb 90-277-0705-7
103. M. Przełęcki, K. Szaniawski and R. Wójcicki (eds.), *Formal Methods in the Methodology of Empirical Sciences.* 1976 ISBN 90-277-0698-0
104. J. M. Vickers, *Belief and Probability.* 1976 ISBN 90-277-0744-8
105. K. H. Wolff, *Surrender and Catch.* Experience and Inquiry Today. [Boston Studies in the Philosophy of Science, Vol. LI] 1976
ISBN 90-277-0758-8; Pb 90-277-0765-0

106. K. Kosík, *Dialectics of the Concrete.* A Study on Problems of Man and World. [Boston Studies in the Philosophy of Science, Vol. LII] 1976
ISBN 90-277-0761-8; Pb 90-277-0764-2
107. N. Goodman, *The Structure of Appearance.* 3rd ed. with an Introduction by G. Hellman. [Boston Studies in the Philosophy of Science, Vol. LIII] 1977
ISBN 90-277-0773-1; Pb 90-277-0774-X
108. K. Ajdukiewicz, *The Scientific World-Perspective and Other Essays, 1931-1963.* Translated from Polish. Edited and with an Introduction by J. Giedymin. 1978
ISBN 90-277-0527-5
109. R. L. Causey, *Unity of Science.* 1977 ISBN 90-277-0779-0
110. R. E. Grandy, *Advanced Logic for Applications.* 1977 ISBN 90-277-0781-2
111. R. P. McArthur, *Tense Logic.* 1976 ISBN 90-277-0697-2
112. L. Lindahl, *Position and Change.* A Study in Law and Logic. Translated from Swedish by P. Needham. 1977 ISBN 90-277-0787-1
113. R. Tuomela, *Dispositions.* 1978 ISBN 90-277-0810-X
114. H. A. Simon, *Models of Discovery and Other Topics in the Methods of Science.* [Boston Studies in the Philosophy of Science, Vol. LIV] 1977
ISBN 90-277-0812-6; Pb 90-277-0858-4
115. R. D. Rosenkrantz, *Inference, Method and Decision.* Towards a Bayesian Philosophy of Science. 1977 ISBN 90-277-0817-7; Pb 90-277-0818-5
116. R. Tuomela, *Human Action and Its Explanation.* A Study on the Philosophical Foundations of Psychology. 1977 ISBN 90-277-0824-X
117. M. Lazerowitz, *The Language of Philosophy.* Freud and Wittgenstein. [Boston Studies in the Philosophy of Science, Vol. LV] 1977
ISBN 90-277-0826-6; Pb 90-277-0862-2
118. Not published
119. J. Pelc (ed.), *Semiotics in Poland, 1894–1969.* Translated from Polish. 1979
ISBN 90-277-0811-8
120. I. Pörn, *Action Theory and Social Science.* Some Formal Models. 1977
ISBN 90-277-0846-0
121. J. Margolis, *Persons and Mind.* The Prospects of Nonreductive Materialism. [Boston Studies in the Philosophy of Science, Vol. LVII] 1977
ISBN 90-277-0854-1; Pb 90-277-0863-0
122. J. Hintikka, I. Niiniluoto, and E. Saarinen (eds.), *Essays on Mathematical and Philosophical Logic.* 1979 ISBN 90-277-0879-7
123. T. A. F. Kuipers, *Studies in Inductive Probability and Rational Expectation.* 1978
ISBN 90-277-0882-7
124. E. Saarinen, R. Hilpinen, I. Niiniluoto and M. P. Hintikka (eds.), *Essays in Honour of Jaakko Hintikka on the Occasion of His 50th Birthday.* 1979
ISBN 90-277-0916-5
125. G. Radnitzky and G. Andersson (eds.), *Progress and Rationality in Science.* [Boston Studies in the Philosophy of Science, Vol. LVIII] 1978
ISBN 90-277-0921-1; Pb 90-277-0922-X
126. P. Mittelstaedt, *Quantum Logic.* 1978 ISBN 90-277-0925-4
127. K. A. Bowen, *Model Theory for Modal Logic.* Kripke Models for Modal Predicate Calculi. 1979 ISBN 90-277-0929-7
128. H. A. Bursen, *Dismantling the Memory Machine.* A Philosophical Investigation of Machine Theories of Memory. 1978 ISBN 90-277-0933-5

129. M. W. Wartofsky, *Models*. Representation and the Scientific Understanding. [Boston Studies in the Philosophy of Science, Vol. XLVIII] 1979
ISBN 90-277-0736-7; Pb 90-277-0947-5
130. D. Ihde, *Technics and Praxis*. A Philosophy of Technology. [Boston Studies in the Philosophy of Science, Vol. XXIV] 1979 ISBN 90-277-0953-X; Pb 90-277-0954-8
131. J. J. Wiatr (ed.), *Polish Essays in the Methodology of the Social Sciences*. [Boston Studies in the Philosophy of Science, Vol. XXIX] 1979
ISBN 90-277-0723-5; Pb 90-277-0956-4
132. W. C. Salmon (ed.), *Hans Reichenbach: Logical Empiricist*. 1979
ISBN 90-277-0958-0
133. P. Bieri, R.-P. Horstmann and L. Krüger (eds.), *Transcendental Arguments in Science*. Essays in Epistemology. 1979 ISBN 90-277-0963-7; Pb 90-277-0964-5
134. M. Marković and G. Petrović (eds.), *Praxis*. Yugoslav Essays in the Philosophy and Methodology of the Social Sciences. [Boston Studies in the Philosophy of Science, Vol. XXXVI] 1979 ISBN 90-277-0727-8; Pb 90-277-0968-8
135. R. Wójcicki, *Topics in the Formal Methodology of Empirical Sciences*. Translated from Polish. 1979 ISBN 90-277-1004-X
136. G. Radnitzky and G. Andersson (eds.), *The Structure and Development of Science*. [Boston Studies in the Philosophy of Science, Vol. LIX] 1979
ISBN 90-277-0994-7; Pb 90-277-0995-5
137. J. C. Webb, *Mechanism, Mentalism and Metamathematics*. An Essay on Finitism. 1980 ISBN 90-277-1046-5
138. D. F. Gustafson and B. L. Tapscott (eds.), *Body, Mind and Method*. Essays in Honor of Virgil C. Aldrich. 1979 ISBN 90-277-1013-9
139. L. Nowak, *The Structure of Idealization*. Towards a Systematic Interpretation of the Marxian Idea of Science. 1980 ISBN 90-277-1014-7
140. C. Perelman, *The New Rhetoric and the Humanities*. Essays on Rhetoric and Its Applications. Translated from French and German. With an Introduction by H. Zyskind. 1979 ISBN 90-277-1018-X; Pb 90-277-1019-8
141. W. Rabinowicz, *Universalizability*. A Study in Morals and Metaphysics. 1979
ISBN 90-277-1020-2
142. C. Perelman, *Justice, Law and Argument*. Essays on Moral and Legal Reasoning. Translated from French and German. With an Introduction by H.J. Berman. 1980
ISBN 90-277-1089-9; Pb 90-277-1090-2
143. S. Kanger and S. Öhman (eds.), *Philosophy and Grammar*. Papers on the Occasion of the Quincentennial of Uppsala University. 1981 ISBN 90-277-1091-0
144. T. Pawlowski, *Concept Formation in the Humanities and the Social Sciences*. 1980
ISBN 90-277-1096-1
145. J. Hintikka, D. Gruender and E. Agazzi (eds.), *Theory Change, Ancient Axiomatics and Galileo's Methodology*. Proceedings of the 1978 Pisa Conference on the History and Philosophy of Science, Volume I. 1981 ISBN 90-277-1126-7
146. J. Hintikka, D. Gruender and E. Agazzi (eds.), *Probabilistic Thinking, Thermodynamics, and the Interaction of the History and Philosophy of Science*. Proceedings of the 1978 Pisa Conference on the History and Philosophy of Science, Volume II. 1981 ISBN 90-277-1127-5
147. U. Mönnich (ed.), *Aspects of Philosophical Logic*. Some Logical Forays into Central Notions of Linguistics and Philosophy. 1981 ISBN 90-277-1201-8
148. D. M. Gabbay, *Semantical Investigations in Heyting's Intuitionistic Logic*. 1981
ISBN 90-277-1202-6

149. E. Agazzi (ed.), *Modern Logic – A Survey.* Historical, Philosophical, and Mathematical Aspects of Modern Logic and Its Applications. 1981 ISBN 90-277-1137-2
150. A. F. Parker-Rhodes, *The Theory of Indistinguishables.* A Search for Explanatory Principles below the Level of Physics. 1981 ISBN 90-277-1214-X
151. J. C. Pitt, *Pictures, Images, and Conceptual Change.* An Analysis of Wilfrid Sellars' Philosophy of Science. 1981 ISBN 90-277-1276-X; Pb 90-277-1277-8
152. R. Hilpinen (ed.), *New Studies in Deontic Logic.* Norms, Actions, and the Foundations of Ethics. 1981 ISBN 90-277-1278-6; Pb 90-277-1346-4
153. C. Dilworth, *Scientific Progress.* A Study Concerning the Nature of the Relation between Successive Scientific Theories. 2nd, rev. and augmented ed., 1986. 3rd rev. ed., 1994 ISBN 0-7923-2487-0; Pallas Pb 0-7923-2488-9
154. D. Woodruff Smith and R. McIntyre, *Husserl and Intentionality.* A Study of Mind, Meaning, and Language. 1982 ISBN 90-277-1392-8; Pb 90-277-1730-3
155. R. J. Nelson, *The Logic of Mind.* 2nd. ed., 1989 ISBN 90-277-2819-4; Pb 90-277-2822-4
156. J. F. A. K. van Benthem, *The Logic of Time.* A Model-Theoretic Investigation into the Varieties of Temporal Ontology, and Temporal Discourse. 1983; 2nd ed., 1991 ISBN 0-7923-1081-0
157. R. Swinburne (ed.), *Space, Time and Causality.* 1983 ISBN 90-277-1437-1
158. E. T. Jaynes, *Papers on Probability, Statistics and Statistical Physics.* Ed. by R. D. Rozenkrantz. 1983 ISBN 90-277-1448-7; Pb (1989) 0-7923-0213-3
159. T. Chapman, *Time: A Philosophical Analysis.* 1982 ISBN 90-277-1465-7
160. E. N. Zalta, *Abstract Objects.* An Introduction to Axiomatic Metaphysics. 1983 ISBN 90-277-1474-6
161. S. Harding and M. B. Hintikka (eds.), *Discovering Reality.* Feminist Perspectives on Epistemology, Metaphysics, Methodology, and Philosophy of Science. 1983 ISBN 90-277-1496-7; Pb 90-277-1538-6
162. M. A. Stewart (ed.), *Law, Morality and Rights.* 1983 ISBN 90-277-1519-X
163. D. Mayr and G. Süssmann (eds.), *Space, Time, and Mechanics.* Basic Structures of a Physical Theory. 1983 ISBN 90-277-1525-4
164. D. Gabbay and F. Guenthner (eds.), *Handbook of Philosophical Logic.* Vol. I: Elements of Classical Logic. 1983 ISBN 90-277-1542-4
165. D. Gabbay and F. Guenthner (eds.), *Handbook of Philosophical Logic.* Vol. II: Extensions of Classical Logic. 1984 ISBN 90-277-1604-8
166. D. Gabbay and F. Guenthner (eds.), *Handbook of Philosophical Logic.* Vol. III: Alternative to Classical Logic. 1986 ISBN 90-277-1605-6
167. D. Gabbay and F. Guenthner (eds.), *Handbook of Philosophical Logic.* Vol. IV: Topics in the Philosophy of Language. 1989 ISBN 90-277-1606-4
168. A. J. I. Jones, *Communication and Meaning.* An Essay in Applied Modal Logic. 1983 ISBN 90-277-1543-2
169. M. Fitting, *Proof Methods for Modal and Intuitionistic Logics.* 1983 ISBN 90-277-1573-4
170. J. Margolis, *Culture and Cultural Entities.* Toward a New Unity of Science. 1984 ISBN 90-277-1574-2
171. R. Tuomela, *A Theory of Social Action.* 1984 ISBN 90-277-1703-6
172. J. J. E. Gracia, E. Rabossi, E. Villanueva and M. Dascal (eds.), *Philosophical Analysis in Latin America.* 1984 ISBN 90-277-1749-4
173. P. Ziff, *Epistemic Analysis.* A Coherence Theory of Knowledge. 1984

ISBN 90-277-1751-7

174. P. Ziff, *Antiaesthetics.* An Appreciation of the Cow with the Subtile Nose. 1984 ISBN 90-277-1773-7
175. W. Balzer, D. A. Pearce, and H.-J. Schmidt (eds.), *Reduction in Science.* Structure, Examples, Philosophical Problems. 1984 ISBN 90-277-1811-3
176. A. Peczenik, L. Lindahl and B. van Roermund (eds.), *Theory of Legal Science.* Proceedings of the Conference on Legal Theory and Philosophy of Science (Lund, Sweden, December 1983). 1984 ISBN 90-277-1834-2
177. I. Niiniluoto, *Is Science Progressive?* 1984 ISBN 90-277-1835-0
178. B. K. Matilal and J. L. Shaw (eds.), *Analytical Philosophy in Comparative Perspective.* Exploratory Essays in Current Theories and Classical Indian Theories of Meaning and Reference. 1985 ISBN 90-277-1870-9
179. P. Kroes, *Time: Its Structure and Role in Physical Theories.* 1985 ISBN 90-277-1894-6
180. J. H. Fetzer, *Sociobiology and Epistemology.* 1985 ISBN 90-277-2005-3; Pb 90-277-2006-1
181. L. Haaparanta and J. Hintikka (eds.), *Frege Synthesized.* Essays on the Philosophical and Foundational Work of Gottlob Frege. 1986 ISBN 90-277-2126-2
182. M. Detlefsen, *Hilbert's Program.* An Essay on Mathematical Instrumentalism. 1986 ISBN 90-277-2151-3
183. J. L. Golden and J. J. Pilotta (eds.), *Practical Reasoning in Human Affairs.* Studies in Honor of Chaim Perelman. 1986 ISBN 90-277-2255-2
184. H. Zandvoort, *Models of Scientific Development and the Case of Nuclear Magnetic Resonance.* 1986 ISBN 90-277-2351-6
185. I. Niiniluoto, *Truthlikeness.* 1987 ISBN 90-277-2354-0
186. W. Balzer, C. U. Moulines and J. D. Sneed, *An Architectonic for Science.* The Structuralist Program. 1987 ISBN 90-277-2403-2
187. D. Pearce, *Roads to Commensurability.* 1987 ISBN 90-277-2414-8
188. L. M. Vaina (ed.), *Matters of Intelligence.* Conceptual Structures in Cognitive Neuroscience. 1987 ISBN 90-277-2460-1
189. H. Siegel, *Relativism Refuted.* A Critique of Contemporary Epistemological Relativism. 1987 ISBN 90-277-2469-5
190. W. Callebaut and R. Pinxten, *Evolutionary Epistemology.* A Multiparadigm Program, with a Complete Evolutionary Epistemology Bibliograph. 1987 ISBN 90-277-2582-9
191. J. Kmita, *Problems in Historical Epistemology.* 1988 ISBN 90-277-2199-8
192. J. H. Fetzer (ed.), *Probability and Causality.* Essays in Honor of Wesley C. Salmon, with an Annotated Bibliography. 1988 ISBN 90-277-2607-8; Pb 1-5560-8052-2
193. A. Donovan, L. Laudan and R. Laudan (eds.), *Scrutinizing Science.* Empirical Studies of Scientific Change. 1988 ISBN 90-277-2608-6
194. H.R. Otto and J.A. Tuedio (eds.), *Perspectives on Mind.* 1988 ISBN 90-277-2640-X
195. D. Batens and J.P. van Bendegem (eds.), *Theory and Experiment.* Recent Insights and New Perspectives on Their Relation. 1988 ISBN 90-277-2645-0
196. J. Österberg, *Self and Others.* A Study of Ethical Egoism. 1988 ISBN 90-277-2648-5
197. D.H. Helman (ed.), *Analogical Reasoning.* Perspectives of Artificial Intelligence, Cognitive Science, and Philosophy. 1988 ISBN 90-277-2711-2

198. J. Wolenski, *Logic and Philosophy in the Lvov-Warsaw School.* 1989 ISBN 90-277-2749-X
199. R. Wójcicki, *Theory of Logical Calculi.* Basic Theory of Consequence Operations. 1988 ISBN 90-277-2785-6
200. J. Hintikka and M.B. Hintikka, *The Logic of Epistemology and the Epistemology of Logic.* Selected Essays. 1989 ISBN 0-7923-0040-8; Pb 0-7923-0041-6
201. E. Agazzi (ed.), *Probability in the Sciences.* 1988 ISBN 90-277-2808-9
202. M. Meyer (ed.), *From Metaphysics to Rhetoric.* 1989 ISBN 90-277-2814-3
203. R.L. Tieszen, *Mathematical Intuition.* Phenomenology and Mathematical Knowledge. 1989 ISBN 0-7923-0131-5
204. A. Melnick, *Space, Time, and Thought in Kant.* 1989 ISBN 0-7923-0135-8
205. D.W. Smith, *The Circle of Acquaintance.* Perception, Consciousness, and Empathy. 1989 ISBN 0-7923-0252-4
206. M.H. Salmon (ed.), *The Philosophy of Logical Mechanism.* Essays in Honor of Arthur W. Burks. With his Responses, and with a Bibliography of Burk's Work. 1990 ISBN 0-7923-0325-3
207. M. Kusch, *Language as Calculus vs. Language as Universal Medium.* A Study in Husserl, Heidegger, and Gadamer. 1989 ISBN 0-7923-0333-4
208. T.C. Meyering, *Historical Roots of Cognitive Science.* The Rise of a Cognitive Theory of Perception from Antiquity to the Nineteenth Century. 1989 ISBN 0-7923-0349-0
209. P. Kosso, *Observability and Observation in Physical Science.* 1989 ISBN 0-7923-0389-X
210. J. Kmita, *Essays on the Theory of Scientific Cognition.* 1990 ISBN 0-7923-0441-1
211. W. Sieg (ed.), *Acting and Reflecting.* The Interdisciplinary Turn in Philosophy. 1990 ISBN 0-7923-0512-4
212. J. Karpiński, *Causality in Sociological Research.* 1990 ISBN 0-7923-0546-9
213. H.A. Lewis (ed.), *Peter Geach: Philosophical Encounters.* 1991 ISBN 0-7923-0823-9
214. M. Ter Hark, *Beyond the Inner and the Outer.* Wittgenstein's Philosophy of Psychology. 1990 ISBN 0-7923-0850-6
215. M. Gosselin, *Nominalism and Contemporary Nominalism.* Ontological and Epistemological Implications of the Work of W.V.O. Quine and of N. Goodman. 1990 ISBN 0-7923-0904-9
216. J.H. Fetzer, D. Shatz and G. Schlesinger (eds.), *Definitions and Definability.* Philosophical Perspectives. 1991 ISBN 0-7923-1046-2
217. E. Agazzi and A. Cordero (eds.), *Philosophy and the Origin and Evolution of the Universe.* 1991 ISBN 0-7923-1322-4
218. M. Kusch, *Foucault's Strata and Fields.* An Investigation into Archaeological and Genealogical Science Studies. 1991 ISBN 0-7923-1462-X
219. C.J. Posy, *Kant's Philosophy of Mathematics.* Modern Essays. 1992 ISBN 0-7923-1495-6
220. G. Van de Vijver, *New Perspectives on Cybernetics.* Self-Organization, Autonomy and Connectionism. 1992 ISBN 0-7923-1519-7
221. J.C. Nyíri, *Tradition and Individuality.* Essays. 1992 ISBN 0-7923-1566-9
222. R. Howell, *Kant's Transcendental Deduction.* An Analysis of Main Themes in His Critical Philosophy. 1992 ISBN 0-7923-1571-5

223. A. García de la Sienra, *The Logical Foundations of the Marxian Theory of Value.* 1992 ISBN 0-7923-1778-5
224. D.S. Shwayder, *Statement and Referent.* An Inquiry into the Foundations of Our Conceptual Order. 1992 ISBN 0-7923-1803-X
225. M. Rosen, *Problems of the Hegelian Dialectic.* Dialectic Reconstructed as a Logic of Human Reality. 1993 ISBN 0-7923-2047-6
226. P. Suppes, *Models and Methods in the Philosophy of Science: Selected Essays.* 1993 ISBN 0-7923-2211-8
227. R. M. Dancy (ed.), *Kant and Critique: New Essays in Honor of W. H. Werkmeister.* 1993 ISBN 0-7923-2244-4
228. J. Woleński (ed.), *Philosophical Logic in Poland.* 1993 ISBN 0-7923-2293-2
229. M. De Rijke (ed.), *Diamonds and Defaults.* Studies in Pure and Applied Intensional Logic. 1993 ISBN 0-7923-2342-4
230. B.K. Matilal and A. Chakrabarti (eds.), *Knowing from Words.* Western and Indian Philosophical Analysis of Understanding and Testimony. 1994 ISBN 0-7923-2345-9
231. S.A. Kleiner, *The Logic of Discovery.* A Theory of the Rationality of Scientific Research. 1993 ISBN 0-7923-2371-8
232. R. Festa, *Optimum Inductive Methods.* A Study in Inductive Probability, Bayesian Statistics, and Verisimilitude. 1993 ISBN 0-7923-2460-9
233. P. Humphreys (ed.), *Patrick Suppes: Scientific Philosopher.* Vol. 1: Probability and Probabilistic Causality. 1994 ISBN 0-7923-2552-4; Set Vols. 1–3 ISBN 0-7923-2554-0
234. P. Humphreys (ed.), *Patrick Suppes: Scientific Philosopher.* Vol. 2: Philosophy of Physics, Theory Structure, and Measurement Theory. 1994 ISBN 0-7923-2553-2; Set Vols. 1–3 ISBN 0-7923-2554-0
235. P. Humphreys (ed.), *Patrick Suppes: Scientific Philosopher.* Vol. 3: Language, Logic, and Psychology. 1994 ISBN 0-7923-2862-0; Set Vols. 1–3 ISBN 0-7923-2554-0
236. D. Prawitz and D. Westerståhl (eds.), *Logic and Philosophy of Science in Uppsala.* Papers from the 9th International Congress of Logic, Methodology, and Philosophy of Science. 1994 ISBN 0-7923-2702-0
237. L. Haaparanta (ed.), *Mind, Meaning and Mathematics.* Essays on the Philosophical Views of Husserl and Frege. 1994 ISBN 0-7923-2703-9
238. J. Hintikka (ed.): *Aspects of Metaphor.* 1994 ISBN 0-7923-2786-1
239. B. McGuinness and G. Oliveri (eds.), *The Philosophy of Michael Dummett.* With Replies from Michael Dummett. 1994 ISBN 0-7923-2804-3
240. D. Jamieson (ed.), *Language, Mind, and Art.* Essays in Appreciation and Analysis, In Honor of Paul Ziff. 1994 ISBN 0-7923-2810-8
241. G. Preyer, F. Siebelt and A. Ulfig (eds.), *Language, Mind and Epistemology.* On Donald Davidson's Philosophy. 1994 ISBN 0-7923-2811-6
242. P. Ehrlich (ed.), *Real Numbers, Generalizations of the Reals, and Theories of Continua.* 1994 ISBN 0-7923-2689-X
243. G. Debrock and M. Hulswit (eds.): *Living Doubt.* Essays concerning the epistemology of Charles Sanders Peirce. 1994 ISBN 0-7923-2898-1
244. J. Srzednicki: *To Know or Not to Know.* Beyond Realism and Anti-Realism. 1994 ISBN 0-7923-2909-0
245. R. Egidi (ed.): *Wittgenstein: Mind and Language.* 1995 (forthcoming) ISBN 0-7923-3171-0

Breinigsville, PA USA
30 December 2009

229710BV00005BA/6/A

9 780792 332459